Context-Aware Machine Learning and Mobile Data Analytics

Iqbal H. Sarker • Alan Colman • Jun Han
Paul Watters

Context-Aware Machine Learning and Mobile Data Analytics

Automated Rule-based Services with Intelligent Decision-Making

Iqbal H. Sarker
Swinburne University of Technology
Melbourne, VIC, Australia

Chittagong University of
Engineering & Technology,
Chittagong, Bangladesh

Jun Han
Swinburne University of Technology
Melbourne, VIC, Australia

Alan Colman
Swinburne University of Technology
Melbourne, VIC, Australia

Paul Watters
Macquarie University
Sydney, NSW, Australia

Cyberstronomy Pty Ltd
Melbourne, Australia

ISBN 978-3-030-88532-8   ISBN 978-3-030-88530-4   (eBook)
https://doi.org/10.1007/978-3-030-88530-4

This Springer imprint is published by the registered company Springer Nature Switzerland AG
The registered company address is: Gewerbestrasse 11, 6330 Cham, Switzerland

*This book is dedicated to my family,
especially my beloved parents, who've
always trusted me and also have encouraged
me to achieve whatever I desired. I would
also like to dedicate this book to my loving
wife and charming son S.M. Irfan Hasan!*

Iqbal H. Sarker, PhD
(First and Corresponding Author)

Preface

We live in the age of "data science and advanced analytics," in which nearly every intelligent service we use in our everyday lives is based on data, collected digitally through smart devices. As the world surrounding an application is continuously updating and computing is heading toward pervasive and ubiquitous environments, *contextual data analytics* are essential for intelligent decision-making and services. Contextual data is information that provides context to an event, person, or object. Smartphones have recently advanced in terms of sensing capabilities, allowing them to collect rich contextual data such as exterior and internal contexts, as well as phone usage records of users in various day-to-day situations. Machine Learning (ML) technology, a core part of Artificial Intelligence (AI), can be used to develop data-driven intelligent context-aware models or systems for smart and automated *decision-making* through extracting insights or useful knowledge, such as rules, from contextual data. Thus, this book provides a thorough knowledge of the notion of context-aware machine learning, as well as automated rule-based modeling in the field of mobile data analytics.

We can divide this book into three main parts:

- At the beginning, we introduce the concept of context-aware machine learning including an automated rule-based framework within the broad area of data science and analytics, particularly, with the aim of data-driven intelligent decision-making. Thus, we have bestowed a comprehensive study on this topic that explores multi-dimensional contexts in machine learning modeling and their usefulness in various context-aware intelligent applications and services.
- In the next part of this book, we present the approaches to extract insights, represented as contextual IF-THEN rules, from mobile data, which can be used for making intelligent decisions in various context-aware test cases. Thus, in this section, we explore various data processing steps and techniques including contextual feature engineering, context discretization with time-series modeling, as well as the approach of extracting contextual IF-THEN rules based on multi-dimensional contexts. Moreover, we also explore the task of recency analysis,

i.e., recent pattern-based updating and management of rules for mobile phone users, which has come to represent an important field of research in the area.
- In the final part of the book, we mainly explore how the contextual information and machine learning rules can be used to make intelligent decisions as well as build various data-driven smart and intelligent applications. We also look into deep learning models, which is a bigger family of machine learning approaches that can be utilized when there is a lot of contextual data. Finally, in terms of new research perspective, future advances in industries or academia, or smart solutions in context-aware technology, prospective research works, and challenges in the field of context-aware computing have been highlighted in this section.

Since a rule-based system has high interpretability and accuracy the automation of discovering rules from contextual raw data can make this book more practical and beneficial for both application developers and researchers. Overall, this book can be used as a useful resource for academics and industry professionals working in various *Fourth Industrial Revolution* (Industry 4.0) application areas, such as Data Science, Machine Learning & AI, Behavioral and Predictive Analytics, Context-Aware Smart Computing, Systems and Personalization, Internet of Things (IoT) and Mobile Applications, as well as Data-Driven Cybersecurity Intelligence with automated rule-based modeling and decision-making.

We are glad to introduce this book to upper-level undergraduate and postgraduate students, as well as academic and industry researchers in the relevant domains mentioned earlier. We would like to express our gratitude to everyone who supported and helped us complete this book. Finally, we would like to express our gratitude to *Springer Nature* for publishing this book. Your insightful feedback on this book would be greatly appreciated.

Enjoy the book!

Melbourne, VIC, Australia	Iqbal H. Sarker
Melbourne, VIC, Australia	Alan Colman
Melbourne, VIC, Australia	Jun Han
Sydney, NSW, Australia	Paul Watters

Acknowledgements

This book would have never been finished without the help of many, to whom I would like to express my sincere thanks. All praise be to the Almighty Allah for providing me the strengths and blessings to complete this book.

I would like to express my sincere gratitude to my PhD supervisors for their exceptional support, patience, and guidance for the completion of this book.

The book's content would not have been possible without the scholarships provided by the Swinburne University of Technology, Melbourne, Australia. I would also like to acknowledge the contribution of the IEEE Computational Intelligence Society (IEEE CIS) in providing IEEE awards, International Conference on Data Science and Advanced Analytics (IEEE DSAA), Canada, that motivated me to work in the area. Finally, I thank everybody who was important to the successful completion of this book with an apology for not mentioning by name.

Contents

About the Authors

Iqbal H. Sarker received his PhD under the Department of Computer Science and Software Engineering at Swinburne University of Technology, Melbourne, Australia, in 2018. Currently, he is working as a faculty member of the Department of Computer Science and Engineering at Chittagong University of Engineering and Technology. His professional and research interests include Data Science, Machine Learning & AI, Data-Driven Cybersecurity and Threat Intelligence, Context-Aware Smart/Intelligence Computing, Smart Cities, Systems and Security. He has published over 100 research papers including top ranked Journals and Conferences with reputed scientific publishers like Elsevier, Springer Nature, IEEE, ACM etc. Recently, he has been listed in the world's TOP 2% Scientist/Researcher list, prepared by Elsevier and Stanford University, USA, 2021. He is one of the founders of the International AIQT Foundation, Switzerland, and a member of ACM and IEEE.
ORCID ID: https://orcid.org/0000-0003-1740-5517.

Alan Colman received his PhD degree in Computer Science and Software Engineering from the Swinburne University of Technology (Melbourne, Australia) in 2006. Since 2006, he has been a researcher and senior lecturer of Software Engineering at the Swinburne University of Technology. His primary research focus has been on adaptive service-oriented software systems, context-aware software systems, and software and Cloud performance prediction and control. He has published over 70 peer-reviewed articles in international journals and conferences.

Jun Han received his PhD degree in Computer Science from the University of Queensland (Brisbane, Australia) in 1992. Since 2003, he has been a professor of Software Engineering at the Swinburne University of Technology (Melbourne, Australia). His primary research focus has been on the architecture and qualities of software systems. His current research interests include dynamic software architectures, context-aware software systems, Cloud and service-oriented software systems, software architecture design, and software performance and security. He has published over 200 peer reviewed articles in international journals and conferences.

Paul A. Watters is Academic Dean at Academies Australasia Polytechnic, an ASX-listed higher education provider operating 18 colleges in Australia and Singapore. Professor Watters is also Honorary Professor of Security Studies and Criminology at Macquarie University, and Adjunct Professor of Cyber Security at La Trobe University. He has worked closely with many large companies and law enforcement agencies in Australia on applied cyber R&D projects, and he has written many books and academic papers on cybersecurity, cybercrime and related topics. His research has been cited 4,964 times, and his h-index is 33. He obtained his PhD at Macquarie University in 2000, and read for his MPhil at the University of Cambridge in 1997 after completing a BA(First Class Honours) at the University of Tasmania, and a BA at the University of Newcastle. Professor Watters is a Fellow of the British Computer Society, a Senior Member of the IEEE, a Chartered IT Professional, and a Member of the Australian Psychological Society.

Part I
Preliminaries

This part of the book consists of the Introduction (Chap. 1), application scenarios and basic structure for context-aware rule learning framework (Chap. 2), and literature review (Chap. 3) to provide the required background knowledge and themes for this book.

Chapter 1
Introduction to Context-Aware Machine Learning and Mobile Data Analytics

1.1 Introduction

Mobile computing and the Internet have played a crucial role in the evolution of the modern digital era. The Internet has firmly established itself as the foundation of modern communication. The use of the Internet, particularly the World Wide Web (WWW), has now spread beyond desktop computers to millions of mobile phones for real-world users.

Mobile devices have become one of the most popular ways for people all over the world to communicate with one another for a variety of reasons. According to the ITU (International Telecommunication Union), cellular network coverage has reached 96.8% of the global population, and in developing countries, this figure reaches almost 100% [1]. Mobile phones have evolved from merely communication devices to intelligent and highly personal essential devices for individual users, capable of assisting them in a variety of daily activities in various day-to-day situations.

In the real world, cell phones can be several types, but in the context of this book, they refer to smartphones or mobile devices with computing and Internet connectivity capabilities. These devices have several important and advanced features that enable better information access through smart computing and proper system utilization for the benefit of the users. Smartphones have become increasingly strong in terms of computing and data storage capacity in recent years. As a result, these smart mobile phones are capable of doing a range of things related to users' everyday lives, such as instant messaging, Internet or web browsing, e-mailing, social network systems, online shopping, or various IoT services like smart cities, health, or transportation services, in addition to being used as a communication device [2, 3]. Future smartphones will be more powerful than current models, capable of communicating faster, storing more data, and incorporating new interaction technologies.

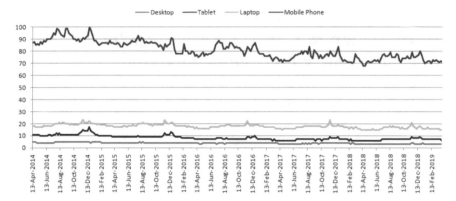

Fig. 1.1 Users' interest trends over time where x-axis represents the timestamp information and y-axis represents the popularity score in a range of 0 (min) to 100 (max)

As a result of recent developments in science and technology around the world, the smartphone sector has seen a fast expansion in the mobile phone application market [4]. According to their diverse capabilities, such as Internet connectivity, data storage, and processing, these devices are well considered as one of the most important Internet-of-Things (IoT) devices [2]. Today's smartphone is also considered as a "next-generation, multi-functional mobile phone that facilitates data processing as well as enhanced wireless connectivity", i.e., a combination of a "powerful cell phone" and a "wireless-enabled PDA" [5]. According to Google Trends data [6], we have shown in our previous paper Sarker et al. [3] that users' interest on *"Mobile Phones"* is growing faster than other platforms like *"Desktop Computer"*, *"Laptop Computer"* or *"Tablet Computer"* for the last 5 years from 2014 to 2019, as shown in Fig. 1.1.

In the real world, people use smartphones for a variety of purposes including e-mailing, instant messaging, online shopping, Internet browsing, entertainment, social media such as Facebook, Linkedin, Twitter, as well as various IoT services such as smart cities, health, and transportation services [2, 3]. The execution environment of smartphone applications differs from that of desktop applications [7]. A desktop computer program is usually created for use in a static execution environment, such as at work or at home. However, in most cases, this static precondition does not apply to mobile networks or systems. The reason for this is that the context in which an application exists is constantly evolving, and computing is heading toward widespread and ubiquitous environments [7]. As a result, mobile applications can adapt to evolving environments and act appropriately, which is known as context-awareness [8].

In the context of computing with smart cell phones, artificial intelligence (AI) techniques have evolved rapidly in recent years, allowing the devices to operate intelligently. In many intelligent mobile applications, such as personalized recommendation, virtual assistant, mobile enterprise, security and privacy, healthcare services, and even the corona-virus COVID-19 pandemic management in recent days, AI-based modeling, especially machine learning techniques [9, 10], and their use in practice can be used. The increasing availability of contextual mobile data and the rapid development of data analytics techniques [11] brought in a paradigm shift to context-aware intelligent computing. Smartphone apps and services are referred to as "context-aware" because they can recognize their users' current contexts and circumstances, "adaptive" because they can change dynamically based on the users' needs, and "intelligent" because they are built on data-driven artificial intelligence, which allows them to assimilate new information and able to assist the end-users intelligently. As a result, machine learning-based modeling for intelligent decision-making is a crucial component of developing context-aware smart mobile apps.

Overall, the aim of this book is to provide a thorough knowledge of the notion of context-aware machine learning, as well as data-driven intelligent decision-making in the field of mobile data analytics. The techniques will aid mobile app developers in creating a variety of data-driven context-aware personalized systems, such as smart context-aware mobile communication, intelligent mobile notification or interruption management, intelligent mobile recommendation, context-aware mobile tourist guide, rule-based predictive modeling, context-aware self-management, and many more that intelligently assist end-users in their daily activities in a pervasive computing environment. Furthermore, the machine learning techniques and decision-making intelligence explored in this book can be applied to a variety of other application domains, such as smart cities and systems, IoT services, cybersecurity & threat intelligence, and many more with relevant data in the area of the Fourth Industrial Revolution (Industry 4.0).

1.2 Context-Aware Machine Learning

In general, machine learning (ML) is considered a branch of Artificial Intelligence (AI), which is based on the idea that systems can learn from data, identify patterns and make decisions with minimal human intervention. ML is closely related to computational statistics, data analytics, or data science, and focuses on computer programs to learn and develop without being explicitly programmed [12]. In other words, machine learning focuses on the development of computer programs that can access data and enable computers to learn and modify behavior without the need for human interaction. The learning process starts with observations or data, such as examples, direct experience, or instructions, to find trends in data and make better decisions in the future. Machine learning models are made up of a collection of rules, methods, or complex transfer functions that can be used to find interesting data patterns or recognize or predict actions [9].

As shown in Fig. 1.2, there are four basic types of machine learning techniques [9]. These are (i) supervised learning that uses labeled data for training algorithms to accurately classify data or predict outcomes; (ii) unsupervised learning is a self-learning strategy in which the system must discover the characteristics of the input population on its own, without the use of a prior collection of categories, and thus

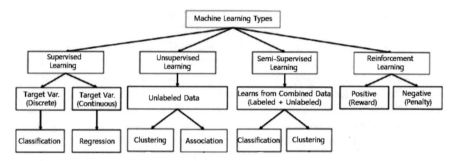

Fig. 1.2 Various types of machine learning techniques

uses unlabeled data to train algorithms; (iii) semi-supervised learning that involves combining a small amount of labeled data with a large amount of unlabeled data during training; and (iv) reinforcement learning, which is concerned with how intelligent agents can behave in a given environment to maximize the concept of cumulative reward. Depending on the nature of the data and the desired outcome, these learning methods can help to build data-driven effective models in a variety of applications.

In this section, we define context-aware machine learning as the learning capabilities from contextual data to build data-driven context-aware systems, particularly intelligent and adaptive applications for smart mobile devices. Thus understanding and defining the contextual information is the first step to go forward. Context has been applied to a variety of fields including mobile and ubiquitous computing, human-centered computing, ambient intelligence, and relevant other areas [13]. Several early research on context-aware computing, or context-awareness, in the field of mobile and ubiquitous computing referred to context as the position of people and things [14]. Schilit et al. [15] argue that the essential aspects of context are (i) where you are, (ii) who you are with, and (iii) what resources are nearby. According to Dey et al. [16] "Context is any information that can be used to characterize the situation of an entity. An entity is a person, place, or object that is considered relevant to the interaction between a user and an application, including the user and the application themselves". We can also define context simply "as a specific type of knowledge to dynamically adapt application behavior according to the current needs", which can be divided broadly into two main categories. These are:

- *External or physical context:* This type of contextual information can be measured by hardware sensors or the device itself. Location, time, light, sound, movement, touch, temperature, air pressure, etc. are some examples of external context.
- *Internal or logical context:* This type of contextual information is mostly specified by the user or captured by monitoring the user's interaction. User's identity, goal, social activity, work context, business processes, preference, emotion, etc. are some examples of logical context.

Table 1.1 Various types of contexts with examples

Context category	Context examples
Temporal context	Date, time of day, season, etc.
Spatial context	Location, orientation, speed, etc.
Social context	People nearby, activity, calendar events, etc.
Environmental context	Temperature, light, noise, etc.
Resource	Nearby, printer, availability, etc.
Computation	CPU, OS, memory, interfaces, etc.
Network	Wire/wireless, bandwidth, error rate, etc.
Physiological	Blood pressure, heart rate, tone of voice, etc.
Psychology	Preference, emotion, tiredness, etc.

Based on the contextual information defined above, context-awareness is the ability of a system or system component to gather such information and adapt behaviors accordingly, which can be the spirit of pervasive computing [17]. As a result, context awareness simply represents the dynamic nature of an application. In a ubiquitous computing environment, the use of contextual knowledge in mobile apps may minimize the amount of human effort and attention required for an application to provide solutions related to a user's needs or preferences [18]. Overall, context-aware systems are concerned with acquiring context (e.g., by utilizing sensors to sense a scenario), understanding and analyzing context (e.g., decision-making according to the context), and implementing application behavior depending on the recognized context (e.g. triggering actions based on context). In Table 1.1, several contexts with examples have been summarized, which might have different impacts on the applications.

In a context-aware system, machine learning techniques can play a significant role to make such decisions dynamically through learning from the contextual data. Thus, this book is based on context-aware machine learning techniques to extract insights or useful knowledge, e.g., rules, from the contextual data, which can be used to build data-driven intelligent context-aware models or systems for smart and automated decision-making in an application.

Typically, a rule $(A \Rightarrow C)$ is any statement that relates two principal components, the rule's left-hand-side (antecedent, A) and the rule's right-hand-side (consequent, C) together. An antecedent states the condition (IF) and consequent states the result (THEN) held from the realization of this condition, i.e., (IF-THEN logical statement). A contextual behavioral rule is defined as $[contexts \Rightarrow behavior]$, where the $contexts$ (antecedent) represents an individual mobile phone user's contextual information (one or many), and the $behavior$ (consequent) represents his/her behavioral activities or usage for that contexts. The reason for using a rule-based method in this book is that rules are easy to understand and can represent the required information efficiently and effectively. Furthermore, rule-based models can be easily expanded to meet specific needs by adding, removing, or modifying them using advanced analysis, such as extracting recent trends or domain experts' expertise.

1.3   Mobile Data Analytics

People nowadays use smartphones for a range of everyday activities including voice communication, text messaging, Internet or web browsing, using a variety of mobile applications (apps), e-mail, social network systems, online messaging, and so on [19]. Individual cell phone users' activities in different contexts can be recorded from various sources such as phone logs, sensors, or other relevant external sources as shown in Fig. 1.3. The data generated by smartphones provide a means of gaining new information about various aspects of the users, such as users' diverse activities with the devices in their daily life, user social interactions, and so on, which allows a better understanding of individual mobile phone users in various contexts in the real world. In this book, we mainly focus on historical phone log data consisting of individual mobile phone users' behavioral activities and corresponding contextual information [20, 21]. Phone call logs [22, 23], SMS Logs [24], Apps usages logs [25, 26], mobile phone notification logs [27], weblogs [28], game Logs [29], context logs [20, 25], and smartphone life logs [30] etc. are some examples of such phone log data. Due to the recent developments in smartphones and their sensing capabilities, the devices can record such rich contextual information about the user as well as device usage records through device logs [25].

The key feature of this type of log data is that it contains the real data of behavioral activities of individual mobile phone users in various situations, as their smartphones automatically record this information based on the users' real-world surroundings. Here, the user's context is defined as "any information that can be used to characterize the situation of the user" (e.g., user current location) [16]. This allows researchers to analyze individual mobile phone user behavior from phone log data and to extract useful knowledge or rules about their actions in various contexts. Individual cell phone users' extracted contextual behavioral rules thus can be used

Fig. 1.3 Various types of data generated by mobile devices

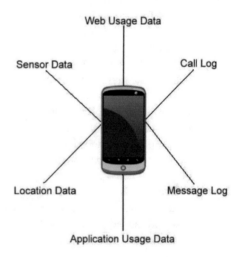

to design and develop adaptive, intelligent mobile applications in a context-aware pervasive computing environment.

Data science methods can help to solve problems by analyzing data and turns data into knowledge [11, 31]. In general, data science is the computing process of discovering patterns in large datasets involving methods at the intersection of machine learning, statistics, and database systems [32]. However, the purpose of mobile data analytics is slightly different, as smart mobile phones are aware of their user's real-life surrounding environment, and users' various types of activities or social interactions, in various contexts in the real world. According to [33], mobile data analytics is the process of intelligently analyzing continuous data streams on mobile devices and can be used as a supporting technology to reduce the cost of collecting user's real-life data and transmission by performing intelligent processing of data for intelligent mobile applications. Thus, mobile data analytics is considered as the area which is concerned with the challenge of finding data-driven models to make dynamic decisions for individual mobile phone users based on relevant contextual information, or simply the contextual patterns and related user actions from the data collected by the smart mobile phones in a pervasive computing environment. In other words, it combines context-aware mobile computing, data mining, machine learning, and a pervasive computing environment. It is the computing process of extracting new knowledge (previously unknown) related to user's activities and their associated contexts, from the mobile phone data, e.g., context logs, which can be used to build real-life smart mobile applications for the end mobile phone users.

Overall, we can conclude that mobile data analytics is a research area focusing on modeling user preferences or habits in different contexts, to analyze various behavioral trends, and ultimately predict behavioral activities from cell phone data, where context-aware machine learning is the key. Thus it can play an important role in a wide range of applications for delivering automated and intelligent services to assist mobile phone users in their everyday lives.

1.4 An Overview of This Book

This book looks mainly at how the useful insights or context-aware machine learning rules are extracted from mobile phone data with examples, and how these rules can be used to design intelligent context-aware applications. It is to be noted that context-aware strategies are different from the general data-driven system in terms of adaptation, intelligence and smartness. Thus, the basic outline of this book has been presented below that covers a background analysis of context-aware machine learning framework with contextual data processing, feature analysis, time-series modeling, rule discovery with multiple contexts, recent pattern analysis, rule-based expert system modeling, the evolution of deep learning-based modeling and future challenges.

This chapter includes an introduction to the definitions, concepts, and principles of context-aware machine learning along with mobile data analytics towards rule-based intelligent applications, where the intentions and goal of this book have been clarified.

Chapter 2 describes the core elements of a generic abstract architecture for context-aware rule-based systems. This architecture serves as a foundation for the examples that follow in the book, with each example incorporating its unique features. We start by presenting application scenarios for two different types of social context-aware applications in this chapter, which motivates research into context-aware machine learning frameworks and systems. This chapter also explores the various stages of this systemic architecture of a context-aware rule-based system, beginning with raw contextual data to services.

Chapter 3 provides the background and reviews the related work from various areas within the scope of this book to analyze and position this work concerning the existing literature. The scope is determined based on the basic architecture mentioned in the previous chapter. It covers the contexts and context-aware computing, the continuous time-series modeling, the rule discovery techniques including association and classification techniques, dynamic rule updating and maintenance techniques including incremental rule mining, and recent log-based rule mining in the related real-life applications for the end mobile phone users. The review also highlights the principal research areas in which the existing solutions fall short of requirements for discovering a concise set of contextual rules of individual mobile phone users, and motivates the research for extracting these rules utilizing smartphone data.

Chapter 4 presents several real-world contextual datasets that can be utilized for experimental purposes to extract useful knowledge or contextual rules. This chapter also represents the contextual feature selection methods for efficient processing.

Chapter 5 presents discretization analysis for time-series behavioral data of smartphone users. Although static segmentation is simple to understand and can be beneficial for analyzing population behavior by comparing across individuals, the generated static segments might not always correspond to individual user activity patterns and subsequent behavior. In this chapter, we focus on dynamic segmentation and modeling by taking into account the diverse behaviors of individuals over time-of-the-week. For this, we describe a behavior-oriented time segmentation methodology that uses phone usage data to generate optimal time segments of individuals with similar behavioral characteristics. This provides a pathway for the extraction of individual mobile phone users' temporal behavioral rules. Finally, how to use the created clusters to generate a set of temporal behavioral rules based on users' preferences for making intelligent decisions in temporal contexts using time-series mobile phone data have been provided.

Chapter 6 focuses on discovering a set of rules based on multi-dimensional contexts—for example, temporal, spatial, social, or other relevant. As a result, we present a rule-based machine learning strategy for extracting a concise set of association rules while accounting for relevant contexts. This chapter also presents how we use a top-down strategy to create an association generation tree based on the relevant multi-dimensional contexts in our rule discovery technique. Finally, how the behavioral association rules are extracted from the tree has been presented.

Chapter 7 presents a recency-based approach, that not only removes the outdated rules from the existing rule-set but also outputs a complete set of updated rules according to individuals' recent behavioral patterns. This chapter presents how we dynamically identify the period for which a recent behavioral pattern has been dominant by analyzing the behavioral characteristics of individual mobile phone users utilizing their mobile phone data. This chapter also presents how to identify the outdated rules from the existing rule-set and outputs a complete set of updated rules from the recent log data.

Chapter 8 explores primarily the mobile expert system, which is considered one of the key AI techniques that can be used to make intelligent decisions and more powerful mobile applications. Hence, we define and explain the mobile expert system as knowledge or rule-based modeling, where a set of context-aware rules are extracted from mobile data using rule-based machine learning methods.

Chapter 9 discusses the importance of deep learning in context-aware behavior modeling for mobile phone users. This chapter also represents a neural network-based deep learning modeling with multiple hidden layers based on the contextual mobile phone data. Finally, this chapter shows the challenges of using deep models along with feasible solutions.

Chapter 10 concludes the book by summarizing several real-world context-aware applications that intelligently assist individual smartphone users in their everyday activities. This chapter also addresses the most important and vital issues, ranging from contextual data collection to decision-making, that have been thoroughly explored. Several prospective research works and challenges in the field of context-aware computing have also been addressed in this chapter.

1.5 Conclusion

In this chapter, we have explored context-aware machine learning related to various aspects to analyze the mobile phone data. Intelligently analyzing the contextual data including both the internal and external contexts with their patterns can be used to build a data-driven context-aware model for smart and automated decision-making, where machine learning technologies are the key. There are different approaches to context-aware machine learning and some glimpses are presented in this chapter. It is to be noted that context-aware machine learning-based strategies are different from basic data-driven applications in terms of adaptation, intelligence and smartness. We have also highlighted the advantages of machine learning rule-based modeling for intelligent decision-making in real-world applications. Overall, it is noted that the machine learning techniques and decision-making intelligence explored in this book can be applied to a variety of application domains such as mobile applications, smart cities and systems, IoT services, cybersecurity intelligence, and many more with relevant data in the area of the Fourth Industrial Revolution (Industry 4.0), mentioned in this chapter.

References

1. Sarker, I. H. (2019). Context-aware rule learning from smartphone data: survey, challenges and future directions. *Journal of Big Data, 6*(1), 1–25.
2. El Khaddar, M. A., & Boulmalf, M. (2017). Smartphone: The ultimate IoT and IoE device. *Smartphones from an Applied Research Perspective, 137.*
3. Sarker, I. H., Hoque, M. M., Uddin, M. K., & Alsanoosy, T. (2021). Mobile data science and intelligent apps: Concepts, AI-based modeling and research directions. *Mobile Networks and Applications, 26*(1), 285–303.
4. Peng, M., Zeng, G., Sun, Z., Huang, J., Wang, H., & Tian, G. (2018). Personalized app recommendation based on app permissions. *World Wide Web, 21*(1), 89–104.
5. Zheng, P., & Ni, L. M. (2006). Spotlight: The rise of the smart phone. *IEEE Distributed Systems Online, 7*(3), 3.
6. Google trends (2019). https://trends.google.com/trends/
7. Finin, T., Joshi, A., Kagal, L., Ratsimore, O., Korolev, V., & Chen, H. (2001, September). Information agents for mobile and embedded devices. In *International workshop on cooperative information agents* (pp. 264–286). Berlin, Heidelberg: Springer.
8. de Almeida, D. R., de Souza Baptista, C., da Silva, E. R., Campelo, C. E., de Figueirêdo, H. F., & Lacerda, Y. A. (2006, April). A context-aware system based on service-oriented architecture. In *20th international conference on advanced information networking and applications-volume 1* (AINA'06) (Vol. 1, pp. 6-pp). IEEE.
9. Sarker, I. H. (2021). Machine learning: Algorithms, real-world applications and research directions. *SN Computer Science, 2*(3), 1–21.
10. Sarker, I. H., Furhad, M. H., & Nowrozy, R. (2021). AI-driven cybersecurity: an overview, security intelligence modeling and research directions. *SN Computer Science, 2*(3), 1–18.
11. Sarker, I. H. (2021). Data science and analytics: An overview from data-driven smart computing, decision-making and applications perspective. *SN Computer Science, 2*, 377
12. Han, J., Kamber, M., & Pei, J. (2011). Data mining concepts and techniques third edition. *The Morgan Kaufmann Series in Data Management Systems, 5*(4), 83–124.
13. Dourish, P. (2004). What we talk about when we talk about context. *Personal and Ubiquitous Computing, 8*(1), 19–30.
14. Schilit, B. N., & Theimer, M. M. (1994). Disseminating active map information to mobile hosts. *IEEE network, 8*(5), 22–32.
15. Schilit, B., Adams, N., & Want, R. (1994, December). Context-aware computing applications. In *1994 first workshop on mobile computing systems and applications* (pp. 85–90). IEEE.
16. Dey, A. K. (2001). Understanding and using context. *Personal and Ubiquitous Computing, 5*(1), 4–7.
17. Shi, Y. (2006, August). Context awareness, the spirit of pervasive computing. In *2006 first international symposium on pervasive computing and applications* (pp. 6–6). IEEE.
18. Anagnostopoulos, C., Tsounis, A., & Hadjiefthymiades, S. (2005, July). Context management in pervasive computing environments. In *ICPS'05. Proceedings. International Conference on Pervasive Services, 2005* (pp. 421–424). IEEE.
19. Pejovic, V., & Musolesi, M. (2014, September). InterruptMe: Designing intelligent prompting mechanisms for pervasive applications. In *Proceedings of the 2014 ACM International Joint Conference on Pervasive and Ubiquitous Computing* (pp. 897–908).
20. Cao, H., Bao, T., Yang, Q., Chen, E., & Tian, J. (2010, October). An effective approach for mining mobile user habits. In *Proceedings of the 19th ACM International Conference on Information and Knowledge Management* (pp. 1677–1680).
21. Hong, J., Suh, E. H., Kim, J., & Kim, S. (2009). Context-aware system for proactive personalized service based on context history. *Expert Systems with Applications, 36*(4), 7448–7457.
22. Phithakkitnukoon, S., Dantu, R., Claxton, R., & Eagle, N. (2011). Behavior-based adaptive call predictor. *ACM Transactions on Autonomous and Adaptive Systems, 6*(3), 1–28.

23. Sarker, I. H., Colman, A., Kabir, M. A., & Han, J. (2016, September). Phone call log as a context source to modeling individual user behavior. In *Proceedings of the 2016 ACM International Joint Conference on Pervasive and Ubiquitous Computing: Adjunct* (pp. 630–634).
24. Eagle, N., & Pentland, A. S. (2006). Reality mining: sensing complex social systems. *Personal and Ubiquitous Computing, 10*(4), 255–268.
25. Zhu, H., Chen, E., Xiong, H., Yu, K., Cao, H., & Tian, J. (2014). Mining mobile user preferences for personalized context-aware recommendation. *ACM Transactions on Intelligent Systems and Technology, 5*(4), 1–27.
26. Srinivasan, V., Moghaddam, S., Mukherji, A., Rachuri, K. K., Xu, C., & Tapia, E. M. (2014, September). Mobileminer: Mining your frequent patterns on your phone. In *Proceedings of the 2014 ACM international joint conference on pervasive and ubiquitous computing* (pp. 389–400).
27. Mehrotra, A., Hendley, R., & Musolesi, M. (2016, September). PrefMiner: Mining user's preferences for intelligent mobile notification management. In *Proceedings of the 2016 ACM International Joint Conference on Pervasive and Ubiquitous Computing* (pp. 1223–1234).
28. Halvey, M., Keane, M. T., & Smyth, B. (2005, September). Time-based segmentation of log data for user navigation prediction in personalization. In *The 2005 IEEE/WIC/ACM international conference on web intelligence (WI'05)* (pp. 636–640). IEEE.
29. Paireekreng, W., Rapeepisarn, K., & Wong, K. W. (2009). Time-based personalised mobile game downloading. In *Transactions on edutainment II* (pp. 59–69). Berlin, Heidelberg: Springer.
30. Rawassizadeh, R., Tomitsch, M., Wac, K., & Tjoa, A. M. (2013). UbiqLog: A generic mobile phone-based life-log framework. *Personal and Ubiquitous Computing, 17*(4), 621–637.
31. Witten, I. H., & Frank, E. (2002). Data mining: Practical machine learning tools and techniques with Java implementations. *ACM SIGMOD Record, 31*(1), 76–77.
32. Cao, L. (2017). Data science: A comprehensive overview. *ACM Computing Surveys, 50*(3), 1–42.
33. Haghighi, P. D., Krishnaswamy, S., Zaslavsky, A., Gaber, M. M., Sinha, A., & Gillick, B. (2013). Open mobile miner: A toolkit for building situation-aware data mining applications. *Journal of Organizational Computing and Electronic Commerce, 23*(3), 224–248.

Chapter 2
Application Scenarios and Basic Structure for Context-Aware Machine Learning Framework

2.1 Motivational Examples with Application Scenarios

In today's world, the increasing adoption and popularity of mobile phones have radically changed the way we connect and communicate with others [1]. The cell phone is a highly personal device that is used in a person's everyday life. These phones are regarded as "always on, always connected" devices [2]. However, due to their day-to-day situations in their everyday lives, cell phone users are not always attentive and responsive to incoming communications. As a result, people are often distracted by incoming phone calls, which not only bother the owners/users but also the people in the close surroundings (Figs. 2.1 and 2.2 illustrate two real-world scenarios). In an official/working environment (e.g., in a meeting/seminar), such interruptions may result in embarrassment, reduce worker performance, increase mistakes, and stress [3]. Furthermore, this can affect other things such as doctoring patients or driving a car, which can result in an accident.

Interruptions consume 28% of a knowledge worker's day, according to the Basex BusinessEdge study [4], which is focused on surveys and interviews conducted by Basex over 18 months with high-level knowledge employees, senior executives at end-user organizations, and executives at companies that manufacture Collaborative Business Knowledge resources. Companies in the United States alone lose 28 billion man-hours per year as a result of this. According to the Bureau of Labor Statistics [5], it results in a loss of $700 billion, based on an average wage of $25/h for a knowledge worker. Bailey et al. [6] found that when users are interrupted, they take 3–27% longer to complete tasks and make twice as many errors. As a result, handling phone call interruptions is crucial, and a context-aware machine learning-based adaptive and intelligent system could be the best option.

Developing machine learning-based computational models capturing user behavioral patterns, analyzing and eventually predict the next behaviors or detect strange behaviors from users' mobile phone data can be used to assist themselves in their

Fig. 2.1 Interruptions with phone calls for the user while driving and may cause car accident

Fig. 2.2 Interruptions with phone calls for the surrounding people and may create embarrassing situations

daily life. The following examples intuitively illustrate the advantage of building such machine learning models.

Motivating Example 1 Let's say Alice is a smartphone user who serves as an executive officer in a corporate office. On Mondays between 9:00 a.m. and 11:00 a.m., she attends a routine meeting at her office. She usually ignores incoming phone calls during that period because she does not want to be distracted during the meeting. The reason for this is that interruptions can not only bother her, but they may also bother others. Even though she is in a meeting, she needs to answer the phone if it is from her boss because it is likely to be important to her. As a result, a mobile phone user's behavior can change according to her current contextual information. Alice's call response behavioral rules can be discovered by a context-aware machine learning model based on her contexts utilizing her mobile phone data, e.g., phone call logs, in which she accepts or declines incoming phone calls in her everyday work routine [7]. The contextual behavioral rules discovered from Alice's phone log data could be used to create a smart call interruption management system for her, providing context-aware customized services to intelligently manage incoming calls.

Fig. 2.3 Various types of smartphone apps that can assist in our daily life activities according to users' current needs

Motivating Example 2 Let's assume Bob, another smartphone user is a Ph.D. candidate. On his smartphone, Bob has a huge range of mobile applications known as apps (Fig. 2.3 illustrates a real-world scenario with various types of apps). Smartphone home screens offer quick access to frequently used applications, which is especially useful [8]. However, Bob's smartphone's home screen is unaware of his changing contexts and as a result, he is unable to intelligently handle applications based on his current needs. Thus a context-aware machine learning model could discover Bob's app usage behavioral rules using his cell phone data, e.g., app usage logs, enabling him to quickly access the app he requires. The contextual behavioral rules discovered from Bob's app usage log data could be used to create a smart mobile app management system that can anticipate his potential usages based on his current contextual information and intelligently assist him in using various types of mobile apps.

The scenarios above demonstrate how a context-aware machine learning model for individual cell phone users, based on data collected by the mobile device, can be a key requirement for developing adaptive, intelligent, and context-aware smart mobile systems that dynamically provide them personalized services in a context-aware ubiquitous computing environment.

2.2   Structure and Elements of Context-Aware Machine Learning Framework

Sensing contextual features, thinking/decision-making, and acting/behaving properly are the three essential functionalities of an intelligent context-aware system. Each of these functionalities involved in a system may differ in complexity. Some systems may have sophisticated sensors for data acquisition, but they need little processing or reasoning before functioning. Others may have little sensing but a huge data processing task is needed before acting for the desired outcome. These systems may also be developed in a centralized or distributed architecture across one or more physical devices in the real world. In the following, we discuss the elements of a context-aware machine learning framework.

Figure 2.4 depicts the basic structure of a context-aware machine-learning architecture highlighting various components ranging from raw contextual data to real-world applications and services, which can make intelligent decisions in a system or application. Based on this architecture, a context-aware system can

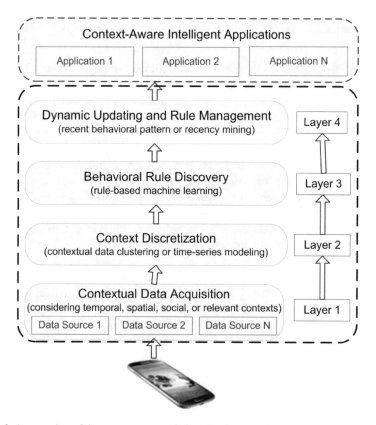

Fig. 2.4 An overview of the context-aware rule learning framework

perform intelligently as well as personalized services as it can discover contextual rules learning from the raw data. As shown in Fig. 2.4, the framework usually consists of four layers: contextual data acquisition layer, context discretization layer, behavioral rule discovery layer, and finally dynamic updating and rule management layer. We will go through these layers and their functions in learning contextual rules from smartphone data in the following. These are:

2.2.1 Contextual Data Acquisition

In general, collecting real-world data is the first step in building a data-driven system [9]. Thus in our context-aware rule learning framework, this is the first layer, which is responsible to collect contextual data from the devices. As individuals' behavior are not static may vary from user to user, this layer is mainly focusing on collecting individual's smartphone data that includes their day-to-day activities with their phones in different contexts such as temporal context, spatial context, social context, and relevant others. Contextual data can be gathered from a variety of sources including phone logs, sensors, and other external sources related to the target application. The key feature of such real data is that it contains the actual behavioral activities of individual mobile phone users in various situations, as their smartphones automatically record this information based on the users' real-world surroundings [1]. Thus, the data generated by smartphones provide a means of gaining new information about various aspects of the users, such as users' diverse activities with the devices in their daily life, user social interactions, and so on, which allows a better understanding of behavioral patterns of the users. To discover such patterns from the raw contextual data, several data processing layers discussed below might be helpful, which are used as the foundation for learning contextual rules and corresponding context-aware applications.

2.2.2 Context Discretization

Context discretization based on machine learning techniques is the second layer in our context-aware rule learning framework, as shown in Fig. 2.4. When we have contextual raw data from the data acquisition layer, we need to discretize the continuous contexts to explain the actual significance of the data, which can also be defined as contextual data clustering. In other words, contextual data with similar characteristics are grouped in one cluster, while data with dissimilar characteristics is grouped in another, and so on. Real-world mobile data incorporates continuous raw contextual information, for example, time-series data, which reflects an individual's varied behaviors in various data points in time order [10]. Although time is the most important context in a mobile Internet portal, such exact time in different time points is not very informative to build a rule-based model. Thus, time-based effective behavior modeling is considered an open problem in the field of context-aware computing. An effective discretization technique, for example, time-series modeling or clustering using machine learning techniques could produce the desired outcome based on the data patterns in the source. Such discretization outcome of a continuous context can be used in the discovery of hidden patterns or relations that serve as the foundation for learning useful contextual rules. Overall, the primary goal of this layer is to present the raw contextual data in a discretized manner based on data characteristics.

2.2.3 Contextual Rule Discovery

The third layer in our context-aware rule learning framework, shown in Fig. 2.4, is machine learning-based contextual rule discovery. Since various contexts can have different effects on people's behavioral activities in the real world, context precedence analysis and corresponding rule discovery can help while making intelligent decisions in various context-aware test cases [7]. This layer is responsible mainly for generating a collection of users' behavioral rules based on the precedence of contexts, taking into account multi-dimensional contexts such as temporal context, spatial context, social context, and other relevant contexts. To discover rules utilizing log dataset, classification rules [11] and association rules [12] are the most common methods, used in the area of mining mobile phone data. However, several issues such as over-fitting problems, redundant generation, and model complexity might happen based on such existing techniques. Thus this could be a major research area presented in this book is to discover a concise set of useful behavioral rules based on multi-dimensional contexts contained in individual's mobile phone data, considering the effectiveness and efficiency of the rule-based system.

2.2.4 Dynamic Updating and Management of Rules

The final layer in our context-aware rule learning system, shown in Fig. 2.4, is machine learning-based dynamic updating and management of the discovered rules. Mobile phone log data is not static as it is progressively added to day-by-day according to an individual's present (on-going) behaviors with mobile phones. Since an individual's behavior changes over time, the most recent patterns, e.g., recency, are more likely to be interesting and significant than older ones for predicting an individual's future behavior in a particular context [13]. Thus recency analysis and mining as well as corresponding rule updation, will play a role in dynamically updating the discovered rules over time, as illustrated in Fig. 2.4. The key advantage of this layer is that it considers the most recent trend, which reflects the freshness of individuals' actions in a given context and is likely to be more important than older patterns in predicting outcomes. As a result, this layer is one of the most important layers in the system, as it is responsible for identifying behavioral trends that evolve, as well as updating and managing rules continuously in response to changes in behavioral patterns.

Overall, the main goal of this machine learning-based framework is extracting a collection of contextual behavioral rules for individual smartphone users from mobile phone data. The extracted rules can be used to create a variety of rule-based intelligent systems that can provide not only the target personalized services that differ from user to user but also population services in the specific application areas.

2.3 Conclusion

We have summarized the basic architecture of a context-aware machine learning system in this chapter. In addition, technologies and techniques used in sensing thought, and acting subsystems have been noted. The main goal of this chapter is to provide a detailed description of the steps involved in processing raw contextual data to build an intelligent system. In the proposed structure, we looked at four different layers of data-driven tasks. We've gone through how to choose the right segmentation strategy, machine learning rule-based modeling, and decision making, as well as the benefits of each layer. The methodologies based on machine learning techniques in applications and scenarios will be discussed in the following chapters, with this architecture serving as the underlying blueprint.

References

1. Sarker, I. H., Hoque, M. M., Uddin, M. K., & Alsanoosy, T. (2021). Mobile data science and intelligent apps: Concepts, AI-based modeling and research directions. *Mobile Networks and Applications, 26*(1), 285–303.
2. Chang, Y. J., & Tang, J. C. (2015, August). Investigating mobile users' ringer mode usage and attentiveness and responsiveness to communication. In *Proceedings of the 17th International Conference on Human-Computer Interaction with Mobile Devices and Services* (pp. 6–15).
3. Pejovic, V., & Musolesi, M. (2014, September). InterruptMe: Designing intelligent prompting mechanisms for pervasive applications. In *Proceedings of the 2014 ACM International Joint Conference on Pervasive and Ubiquitous Computing* (pp. 897–908).
4. Spira, J. B., & Feintuch, J. B. (2005). The cost of not paying attention: How interruptions impact knowledge worker productivity. Report from Basex.
5. Bureau of Labor Statistics. http://www.bls.gov
6. Bailey, B. P., & Konstan, J. A. (2006). On the need for attention-aware systems: Measuring effects of interruption on task performance, error rate, and affective state. *Computers in Human Behavior, 22*(4), 685–708.
7. Sarker, I. H., & Kayes, A. S. M. (2020). ABC-RuleMiner: User behavioral rule-based machine learning method for context-aware intelligent services. *Journal of Network and Computer Applications, 168*, 102762.
8. Sarker, I. H., & Salah, K. (2019). Appspred: Predicting context-aware smartphone apps using random forest learning. *Internet of Things, 8*, 100106.

9. Sarker, I. H., Furhad, M. H., & Nowrozy, R. (2021). AI-driven cybersecurity: An overview, security intelligence modeling and research directions. *SN Computer Science, 2*(3), 1–18.
10. Sarker, I. H., Colman, A., Kabir, M. A., & Han, J. (2018). Individualized time-series segmentation for mining mobile phone user behavior. *The Computer Journal, 61*(3), 349–368.
11. Quinlan, J. R. (2014). *C4. 5: Programs for machine learning*. Elsevier.
12. Agrawal, R., & Srikant, R. (1994, September). Fast algorithms for mining association rules. In *Proc. 20th Int. Conf. Very Large Data Bases, VLDB* (Vol. 1215, pp. 487–499).
13. Sarker, I. H., Colman, A., & Han, J. (2019). Recencyminer: Mining recency-based personalized behavior from contextual smartphone data. *Journal of Big Data, 6*(1), 1–21.

Chapter 3
A Literature Review on Context-Aware Machine Learning and Mobile Data Analytics

3.1 Contextual Information

The term context has a wide range of meanings and can be applied to a variety of situations. In this section, we first go through some of the existing context definitions in the domain of mobile and pervasive computing, and then we go over why contexts are important in a particular application.

3.1.1 Definitions of Contexts

Context has been employed in a variety of fields, including pervasive and ubiquitous computing, human-computer interaction, computer-supported collaborative work, and ambient intelligence [1]. Early efforts on context-awareness in the area of ubiquitous and pervasive computing referred to context as essentially the location of people and objects [2]. Context has recently been expanded to encompass a broader set of factors, such as an entity's physical and social features, as well as user behaviors [1]. Following a review of the pervasive and ubiquitous computing community's definitions and categories of context, this part aims to describe the concept of the context within the area. Because the concepts of context in the domain of pervasive and ubiquitous computing are similarly broad and this discussion is meant to be informative rather than comprehensive.

From various viewpoints, several research has sought to define and describe the context. Schilit et al. [2], for example, consider the user's location information, the surrounding persons and objects, and the changes to those objects as contexts. Contexts are also defined by Brown et al. [3] as the user's locational information, temporal information, the surrounding individuals around the user, temperature, and so on. In the same way, the user's locational information, ambient information, temporal information, and identity are all considered contexts By Ryan et al. [4].

I. H. Sarker et al., *Context-Aware Machine Learning and Mobile Data Analytics*, https://doi.org/10.1007/978-3-030-88530-4_3

Other context definitions have merely provided synonyms for contexts, such as context as the environment or social condition. A lot of studies have considered the context as the user's environmental information. For example, in [5], Brown et al regarding the environmental information that the user's computer is aware of as context, whereas Franklin et al [6] regards the user's social setting as context. Other researchers, on the other hand, believe it is the environment that is related to the applications. Ward et al. [7], for example, consider the state of the applications' surrounding information as contexts. Context is defined by Hull et al. [8] as the features of the user's current position, which includes the complete surroundings. In Rodden et al. [9], the settings of apps are likewise considered as context.

Schilit et al. [10] argued that the best parts of context are (i) where you are, (ii) who you are with, and (iii) what resources are nearby. In their definition, information about the changing environment is taken into consideration as context. They encompass the computational environment as well as the physical environment, in addition to the user environment, e.g., user location, adjacent individuals, and the current social position of the user. For example, the computing environment can include connection, available processors, user input and display, network capacity, and computing costs, while the physical environment can include noise, temperature, and lighting levels.

Dey et al. [11] give a survey of different views of context, which are mostly imprecise and indirect, often defining context by synonym or example. Finally, he provides the following definition of context, which is now widely accepted. According to Dey et al. [11] "Context is any information that can be used to characterize the situation of an entity. An entity is a person, place, or object that is considered relevant to the interaction between a user and an application, including the user and the application themselves".

3.1.2   Understanding the Relevancy of Contexts

Realizing the importance of contexts is merely the first step in properly utilizing them in mining contextual behavioral rules of individual mobile phone users. We need a clear understanding of what circumstances influence users to make decisions in different situations to make efficient use of contexts in mobile phone users' behavioral rules. The contexts associated with the user are the most relevant as we aim to discover the user behavioral rules using their mobile phone data. Table 3.1 depicts an example of user situations influencing decision-making when dealing with phone call interruptions. The relevance of the contexts, on the other hand, is application particular, i.e., it may differ from one application to the next in the real world.

Consider a personalized smart mobile app management system that can predict an individual's future app usages (Skype, Whatsapp, Facebook, Gmail, Microsoft Outlook) based on contextual data. When the user is in her office on weekdays

Table 3.1 Various types of user contexts influencing making decisions while handling the phone call interruptions

Context category	Context examples
Temporal context	User's activity occuring date (YYYY-MM-DD), time (hh:mm:ss), period (e.g., 1 h, 10:00 a.m.–12:00 p.m.), weekday (e.g., Monday), weekend (e.g., Saturday), etc.
Spatial context	User's coarse level location such as office, work, home, market, restaurant, vehicle, playground etc.
Social context	User's social activity or situation such as professional meeting, lecture, seminar, lunch break, dinner, etc., and/or, social relationship between individuals such as mother, friend, colleague, boss, significant one, unknown, etc.

between 9:00 a.m. and 10:00 a.m., she normally uses Microsoft Outlook for mailing purposes. The user's contexts, such as temporal (Weekdays between 09:00 a.m. and 10:00 a.m.) and place (office), may be relevant to intelligently assist her in finding this particular mobile application among a large number of installed apps on her mobile phone.

Consider another example: a smartphone call interruption management system, which may require more contexts. Mobile phones are commonly considered to being "always on, always connected" devices in the real world, yet mobile users are not always attentive and receptive to incoming contact [12]. Let's say a user has a regular meeting at her office on Monday between 9:00 and 11:00 a.m. She usually rejects incoming phone calls during that period since she does not want to be interrupted during the meeting. If the phone call is from her boss or mother, she wants to answer it since it seems to be important to her. According to this example, user phone call response behaviors are related not only to contexts, location (e.g., workplace), and temporal (e.g., Monday, between 9:00 a.m. and 11:00 a.m.), but also to additional contexts, social situations (e.g., meeting), and social relationships between individuals (e.g., boss or mother). As a result, the relevance of user circumstances differs from app to app in the real world.

With a better understanding of contexts, mobile app developers will be able to choose which contexts to be included in their apps, allowing them to create context-aware apps that deliver personalized services and intelligently aid users in their daily activities as well as smartphone based IoT services [123, 124]. According to the aforementioned real-world examples, individual mobile phone users' behavioral rules should not be dependent on a fixed number of contexts. To meet these needs, we provide a set of behavioral rules for individual mobile phone users based on multi-dimensional contexts available in the mobile phone dataset, which may be employed in relevant applications for the mobile phone user.

3.2   Context Discretization

The discretization of continuous contextual data is one of the key research areas covered in this book. The discretization method converts continuous numerical attributes into discrete or nominal attributes with a finite number of intervals, resulting in a non-overlapping partition of a continuous domain.

3.2.1   Discretization of Time-Series Data

Temporal context, represented as time-series data, is the most important aspect that influences user behavior in a mobile Internet portal [13], according to the scope of continuous context considered in this book. A time series is defined as "a sequence of data points ordered in time, often measured at successive time points" [14]. In this section, we focus on the discretization of time-series data as context discretization.

Unlike digital systems, human perception of time is not precise. Routine behaviors always have a time interval, even if it is only a little one, such as 5 min. Time must be segmented into meaningful categories that act as a proxy for distinguishing user's various activities to evaluate time as a condition in a high confidence rule. As a result, discretization of time-series data is required, which may then be used as the foundation for a mobile phone-based context-aware rule learning system. Its major purpose is to convert continuous time-series attributes into discrete or categorical values, such as time segments, hence converting quantitative data into qualitative data. According to [15], time-based behavior modeling is an open problem. Hence, we summarize the existing time-series segmentation approaches into two broad categories; (i) static segmentation, and (ii) dynamic segmentation, which is used in various mobile applications. In the following, we discuss these methods used in various application domains.

3.2.2   Static Segmentation

A static segmentation is simple to comprehend and can be useful for comparing population behavior among cell phone users. Most researchers recently consider only the temporal coverage (24-hours-a-day) and statically segment time into arbitrary categories (e.g., morning) or times (e.g., 1 h) to produce segments, as shown in Table 3.2. This form of static time segmentation focuses primarily on time intervals. According to [16], there are two forms of time intervals: equal and unequal time intervals.

A number of researchers have used equal interval-based segmentation in their applications. For instance, Song et al. [17] present a log-based analysis on users' search activity in order to increase search relevance by splitting the 24-h day

Table 3.2 Various types of static time segments used in different applications

Time interval type	Number of segments	Used time interval and segment details	References
Equal	3	Morning [7:00–12:00], afternoon [13:00–18:00] and evening [19:00–24:00]	Song et al. [17]
Equal	3	[0:00–7:59], [8:00–15:59] and [16:00–23:59]	Rawassizadeh et al. [18]
Equal	4	Morning [6:00–12:00], afternoon [12:00–18:00], evening [18:00–24:00] and night [0:00–6:00]	Mukherji et al. [19]
Equal	4	Morning [6:00–12:00], afternoon [12:00–18:00], evening [18:00–24:00] and night [0:00–6:00]	Bayir et al. [20]
Equal	4	Morning, afternoon, evening and night	Paireekreng et al. [21]
Equal	4	Morning [6:00–11:59], day [12:00–17:59], evening [18:00–23:59], overnight [0:00–5:59]	Jayarajah et al. [22]
Equal	4	Night [0:00 a.m.-6:00 a.m.], morning [6:00 a.m.-12:00 p.m.], afternoon [12:00 p.m.-6:00 p.m.], and evening [6:00 p.m.-0:00 a.m.]	Do et al. [23]
Unequal	3	Morning (beginning at 6:00 a.m. and ending at noon), afternoon (ending at 6:00 p.m.), night (all remaining hours)	Xu et al. [24]
Unequal	4	Morning [6:00–12:00], afternoon [12:00–16:00], evening [16:00–20:00] and night [20:00–24:00 and 0:00–6:00]	Mehrotra et al. [25]
Unequal	5	Morning [7:00–11:00], noon [11:00–14:00], afternoon [14:00–18:00] and so on	Zhu et al. [26]
Unequal	5	Morning, forenoon, afternoon, evening, and night	Oulasvirta et al. [27]
Unequal	5	Morning [7:00–11:00], noon [11:00–14:00], afternoon [14:00–18:00], evening [18:00–21:00], and night [21:00-Next day 7:00]	Yu et al. [28]
Unequal	>5	Early morning, morning, late morning, midnight and so on	Naboulsi et al. [29]
Unequal	>5	Early morning, morning, late morning, midnight and so on	Dashdorj et al. [30]
Unequal	>5	Early morning, morning, late morning, midnight and so on	Shin et al. [31]
Unequal	8	S1[0:00 a.m.–7:00 a.m.], S2[7:00 a.m.–9:00 a.m.], S3[9:00 a.m.–11:00 a.m.], S4[11:00 a.m.–2:00 p.m.], S5[2:00 p.m.–5:00 p.m.], S6[5:00 p.m.–7:00 p.m.], S7[7:00 p.m.–9:00 p.m.] and S8[9:00 p.m.–12:00 a.m.]	Farrahi et al. [33]

into three equivalent time segments, e.g., morning [7:00–12:00], afternoon [13:00–18:00], and evening [19:00–2400]. Using three temporal segments [0:00–7:59], [8:00–15:59], and [16:00–23:59], Rawassizadeh et al. [18] propose a scalable method for regular behavioral pattern mining from multiple sensor data. Morning [6:00–12:00], afternoon [12:00–18:00], evening [18:00–2400], and night [0:00–6:00] are the four period segments considered by Mukherji et al. [19]. Using the same four time segments, Bayir et al. [20] suggest a web-based customized mobility service for mobile applications. Paireekreng et al. [21] introduced a personalization mobile game recommendation framework using time-of-day divided into four cycles—morning, midday, evening, and night. Jayarajah et al. [22] use morning [6:00–11:59], day [12:00–17:59], evening [18:00–23:59], and overnight [0:00–5:59] to understand the difference in variety seeking over various time windows. In their application model, Do et al. [23] night [0:00 a.m.–6:00 a.m.], morning [6:00 a.m.–12:00 p.m.], afternoon [12:00 p.m.–6:00 p.m.], and evening [6:00 p.m.–0:00 a.m.] to explain how user behavior changes with respect to time of day.

Several researchers have used unequal interval-based segmentation in their applications. For instance, Xu et al. [24] have provided a prediction system for smartphone app usages that incorporates three important everyday factors that affect user app use behavior (context, group behavior, and user preferences). Morning (starting at 6:00 a.m. and finishing at noon), afternoon (ending at 6:00 p.m.), and night (all remaining hours) are the time segments they use. Mehrotra et al. propose a novel interruptibility management solution in [25] that learns users' preferences for receiving mobile alerts based on automated rule extraction by mining their contact with mobile phones. Morning [6:00–12:00], afternoon [12:00–16:00], evening [16:00–20:00], and night [20:00–24:00 and 0:00–6:00] are the four-time slots they use for segmentation. In their recommendation scheme, Zhu et al. [26] use five static time segments in a day that are predefined as morning [7:00–11:00], noon [11:00–1400], afternoon [14:00–18:00], and so on. Oulasvirta et al. [27] use five-time slots (morning, forenoon, afternoon, evening, and night) as temporal context to explain each user's thoughts, ideas, beliefs, and emotions. Yu et al. investigate how to mine topic models to manipulate user context logs for customized context-aware suggestion in [28]. Morning [7:00–11:00], noon [11:00–14:00], afternoon [14:00–18:00], evening [18:00–21:00], and night [21:00-Next day 7:00] are the period segments used throughout their framework.

A number of authors [29–31] add to the above segmentations by introducing early morning, late morning, midnight, and so on. Shin et al. propose a new context model for app prediction in [32], which gathers a wide variety of contextual information in a smartphone and makes customized app predictions using a naive Bayes model. They divide time for weekdays and weekends into early morning, morning, afternoon, evening, and night in their model. Farrahi et al. [33] divide each day into 8 coarse-grain time slots as follows: [0:00 a.m.–7:00 a.m.], [7:00 a.m.–9:00 a.m.], [9:00 a.m.–11:00 a.m.], [11:00 a.m.–2:00 p.m.], [2:00 p.m.–5:00 p.m.], [5:00 p.m.–7:00 p.m.], [7:00 p.m.–9:00 p.m.] and [9:00 p.m.–12:00 a.m.]. These time slots were chosen to represent popular activities in everyday life, such as lunch, dinner, or work hours in the morning and afternoon. These types of segmentation are often used in a variety of applications such as mining mobile user habits [34, 35], managing mobile intelligent interruption management system [36], mining frequent co-occurrence patterns on the mobile phones [37], making app prefetch practical on mobile phones [38].

Several authors use time segments for different events scheduled in their calendar in addition to the above time segments to predict individual cell phone user activity. Cell phones are also considered one of the main means of accessing calendars (e.g., Google Calendar) to coordinate schedules such as meetings since they are still associated and carrying with the users [39]. The calendar [40] allows the user to identify unique tasks or events with length, temporal domain, and other attributes. For example, if the calendar shows a meeting between 13:00 and 14:00, they presume the user is inaccessible and is in a location with at least one other person [41]. The time interval [13:00–14:00] is then used to forecast her cell phone use. Calendar entries, according to Khalil et al. [42], is a good indicator of whether an individual is available or unavailable for a phone call. Salovaara et al. [43] conducted a study and found that 31% of incoming phone calls were due to unavailability, i.e., users were unable to answer the phone calls due to meetings, classes, appointments, driving, or sleeping.

Several authors have designed a context-aware interruption management framework that produces as an output if an incoming call should be enabled to ring by taking into account the user's above unavailability solution for a specific time segment (e.g., between 13:00 and 14:00) using an individual's calendar details. Dekel et al. [44] build an application to reduce cell phone disturbances, for example. The developers of [45] and [36] use calendar information to create a context-aware interruption management system. To enhance mobile phone understanding, Seo et al. [46] use the user's schedule to determine policy rules in their context-aware phone configuration management framework. The interruption handling rules in these methods are focused on static temporal segments based on their scheduled appointments in their individual's calendar details, for example, the user is unable to answer the incoming call while s/he is in a calendar case (e.g., a meeting between 13:00 and 14:00). However, in some situations, such an unavailability approach offers poor accuracy.

Khalil et al. [47] surveyed 72 phone users and discovered that the above unavailability solution for mobile communication has low accuracy (62%) for loosely organized home activities like lunch, watching TV, and doing homework, but high accuracy (93%) for structured events like classes, meetings, and appointments. However, such special terms are insufficient to cover real-world use cases; a larger range of meeting categories keywords is needed to capture users' actual actions [44]. Even if the user is involved in an ongoing task or social situation, the phone call is always not disruptive, and the call is welcomed because it offers a required mental break from the current task [48]. According to [49], 24% of mobile phone users feel compelled to pick up a phone call while in a meeting. According to a user survey conducted by Rosenthal et al. [50], 35% of participants want to receive phone calls at work, while the rest do not. Sarker et al. [41] have demonstrated that the presence of a calendar event for a specific time segment is insufficient to assume individual actions for their different calendar events. According to [51], the calendar does not offer a reliably accurate depiction of the real world because events do not occur or occur outside the calendar's allocated time window. As a result, calendar-based temporal segments are ineffective at capturing the actions of individual mobile phone users [41].

Although different time intervals and corresponding segmentation are used for different purposes (see Table 3.2), these methods all take into account a fixed number of segments for all users. However, users' behavioral evidence that varies from user to user over time in the real world is not taken into account when doing such segmentation. As a result, static segment generation may not be appropriate for producing high-confidence temporal rules for individual smartphone users. For example, in one case, a N_1 number of segments may yield meaningful results, whereas in another case, a N_2 number of segments may yield better results, where $N_1 \neq N_2$. As a result, rather than statically generating rules, dynamic segmentation of time may be able to represent individuals' behavioral evidence over time and play a role in producing high confidence rules based on their utilization records.

3.2.3 Dynamic Segmentation

A segmentation technique that produces a variable number of segments, as discussed above, will be more useful for modeling users' behavior. To achieve the target, a dynamic segmentation technique rather than a static segmentation technique may be used. The number of segments in a dynamic segmentation is not set and predefined; it can change based on behavioral features, patterns, or preferences. There are many dynamic segmentation strategies for modeling users' behavioral patterns in temporal contexts that generate a variable number of segments. To produce the segments, several authors simply take into account a single parameter, such as interval length or base time. Depending on the time frame, the number of time segments varies. The number of segments will be T_{max}/BP [52] if T_{max} reflects the entire 24-h time span and BP is a base period. The number of time segments decreases as the base period increases and vice versa. If the base time is 5 min, the number of segments will be determined by dividing $24\text{-}hours\text{-}a\text{-}day$ by 5. In this case, a base period of 5 min is assumed to be the finest granularity for distinguishing an individual's day-to-day activities. The number of segments decreases as the base time is increased to 15 min, with 15 min being considered to be the finest granularity. As a result, the number of segments varies depending on the starting time frame.

For example, Ozer et al. [53] suggest using sequential pattern mining techniques to predict the location and time of cell phone users. In their process, they use a 15-min time interval for segmentation and then switch to 60-min intervals in their experiments. Do et al. present a system for predicting where users will go and which app they will use next using rich contextual knowledge from smartphone sensors in [54]. They use 30 min as the parameter value in their system. Farrahi et al. use temporal data to discover everyday habits from large-scale cell phone data in [55]. They also divide each day of the week into 30-min segments using 30 min as the parameter value. Karatzoglou et al. use 2-h as the parameter value in their mobile app recommendation system in [56]. In their analysis to classify human daily activity patterns using cell phone data, Phithakkitnukoon et al. [57] use 3 h for time segmentation. For various purposes, different authors use different interval values to create a variable number of segments. When the value of such an interval is high, it results in a small number of segments, and vice versa. However, the optimal value of this parameter, which we are interested in, is required for an effective segmentation that captures individual mobile user behavior.

Individual calendar schedules and corresponding time boundaries may also be used to evaluate variable duration time segments to model users' actions in a temporal sense, which may differ depending on users' preferences [41]. For example, one user may have an event between 1 and 2 p.m., while another may have an event between 1:30 pm and 2:30 pm. As a result, the time segmentation varies depending on the events they have planned in their calendars. Multiple thresholds, sliding windows, and data shape-based methods, as shown in Table 3.3, are also used in many applications. Halvey et al. [58] proposed a multi-thresholds-based approach for segmenting time-series log data to predict mobile device navigation patterns. However, since no previous awareness of user behaviors exists, choosing these thresholds to define the lower and upper boundary of a segment is extremely difficult.

Several authors use machine learning methods such as clustering, genetic algorithms, and others in addition to these approaches. To discover rules from time series, Das et al. [59] suggest a cluster-based technique. The issue is that the number of clusters must be known ahead of time, which is difficult to predict for an individual. Besides these, GA based [60, 61], sliding window-based [62, 63], shape-based [16, 64] segmentation have been proposed for different purposes.

The user's total number of activity occurrences at each time point is used to segment the data. These are not, however, behavior-oriented segmentations since they do not account for the various actions of individuals that we are interested in. Using cell phone data, a variety of authors examine various usage habits over time. Phithakkitnukoon et al. [65], for example, create a behavior-based adaptive call prediction system based on mobile phone data. Jang et al. have shown in [66] that different users' app usage activity differs over time in a day while using mobile data. Henze et al. use mobile phone data in [67] to determine the best time to deploy applications. Xu et al. [68] use cell phone data to determine the best period for active applications. Based on user activity, Bohmer et al. [69] describe the peak time of typical app usages. These methods consider scanning over each hour time slot of the day (for example, [1:00 p.m.–2:00 p.m.]) to capture user habits and locate a specific predefined section for their purposes. Such methods, on the other hand, ignore the complex optimal segmentation based on an individual's actions. We have summarized a variety of works that use dynamic segmentation techniques for various purposes in Table 3.3.

Table 3.3 Various types of dynamic time segments used in different applications

Base technique	Description	References
Single parameter	A predefined value of time interval, e.g., 15 min is used to generate segments	Ozer et al. [53]
	A different value of time interval, e.g., 30 min is used for segmentation	Do et al. [54], Farrahi et al. [55]
	A relatively large value of the parameter, e.g., 2-h is used to generate time segments	Karatzoglou et al. [56]
	Another large value of time interval, e.g., 3-h is used for segmentation to make the number of segments small	Phithakkitnukoon et al. [57]
Calendar	Various calendar schedules and corresponding time boundaries are used to model users' behavior in temporal context	Khail et al. [47], Dekel et al. [44], Zulkernain et al. [36], Seo et al. [46], Sarker et al. [41]
Multi-thresholds	To identify the lower and upper boundary of a segment for segmenting time-series log data	Halvey et al. [58]
Data shape	A data shape based time-series data analysis	Zhang et al. [16], Shokoohi et al. [64]
Sliding window	A sliding window is used to analyze time-series data	Hartono et al. [62], Keogh et al. [63]
Clustering	A predefined number of clusters is used to discover rules from time-series data	Das et al. [59]
Genetic algorithm	A genetic algorithm is used to analyze time-series data	Lu et al. [60], Kandasamy et al. [61]

Clustering, as shown in Table 3.3, is an effective machine learning technique for forming broad time segments that take into account such user activity patterns. Clustering algorithms are typically built on certain assumptions and are biased against certain types of problems. In this sense, saying "best" in the context of clustering algorithms is a challenging task; it depends on the particular application [70]. The K-means algorithm is the most well-known squared error-based clustering algorithm [71] among a variety of clustering algorithms in the area of machine learning and data science. However, this algorithm requires the initial partitions and a fixed number of clusters K to be defined. With different starting points, the convergence centroids often change. Because of the estimation of mean values, outliers may often affect this algorithm. More significantly, this algorithm's characteristics aren't directly applicable to our context-aware rule learning. This algorithm, for example, assigns objects to the nearest cluster using the Euclidean distance function as a measure of similarity. However, Euclidean distance is ineffective for determining individual behavioral similarity and, as a result, learning behavioral rules. In the presence of outliers, another K-medoids method [72], is more robust than the K-means algorithm since a medoid is less affected by outliers than a mean. As it reduces the outlier problem, K-means and the problem of time-series modeling have other characteristics in common.

Since the size and number of time segments are determined by the user's actions, which vary from one user to the next, bottom-up hierarchical data processing may aid in the formation of behavioral clusters. There are two types of hierarchical algorithms currently available: agglomerative methods and divisive methods. The system clustering process, on the other hand, is not widely used in practice [70]. Single linkage [73] and full linkage [74] are the simplest and most common agglomerative clustering methods. The single linkage agglomerative clustering algorithm is similar to another tool, nearest neighbor [70]. All of these hierarchical algorithms rely on a proximity matrix, which is computed by calculating the distance between two new clusters. The clusters are then successively merged according to the matrix value until the desired cluster structure is obtained. Because of the differences in user behavior, it is impossible to predict the degree to which merging is optimal according to a proximity matrix. Thus, using such clustering techniques, segments could be produced based on time-series data on user behavior patterns. Similarly, approaches based on genetic algorithms, such as those shown in Table 3.3, generate dynamic segments.

In summary, time-series modeling, using both the static and dynamic segmentation methods discussed above, can produce a variety of time segments that can be used for a variety of purposes. The above time-series modeling approaches, on the other hand, do not always map to trends of individual users based on their preferences, which are based on users' diverse habits through time-of-week and may differ from user to user. To effectively use temporal context as the basis for discovering rules capturing smartphone user behavior, a machine learning-based time-series modeling technique that takes into account such patterns may be important.

3.3 Rule Discovery

Another major focus of this study is using smartphone data to discover useful behavioral rules of individual cell phone users based on multi-dimensional contexts, such as temporal, spatial, or social contexts. In the field of machine learning, the most popular techniques for discovering such rules of individual cell phone users are association rule learning [75] and classification rule learning [76]. We will provide a brief overview of both association and classification strategies for discovering rules based on multi-dimensional contexts in the following sections.

3.3.1   Association Rule Mining

Association rule mining [77] is the discovery of associations or patterns or rules or relationships among a set of available items in a given dataset. Due to the descriptive and easily understandable existence of the discovered association rules, association rule mining has become a popular data mining technique [77]. Initial research into mining association rules was largely motivated by the analysis of retail market basket data to understand the purchasing behavior of the customers. One example is that "if a customer buys a computer or laptop (an item), s/he is likely to also buy anti-virus software (another item) at the same time". A common way of measuring the usefulness of association rules is to use its parameter, the 'support' and 'confidence' which is introduced in [77]. Support of a rule $Sup(A \Rightarrow C)$ is the percentage (%) of records in the dataset which carries all the items or contexts in a rule, and the confidence $Conf(A \Rightarrow C)$ is the percentage (%) of the records that carry all the items or contexts in the rule among those records that carry the items in the antecedent (A) of the rule.

Association rule mining algorithm discovers association rules that satisfy the predefined minimum support and confidence constraints from a given dataset [75]. The association rule mining problem is usually decomposed into two subproblems; (i) the first one is to identify several item sets whose occurrences exceed the predefined minimum support threshold in the dataset, those item sets are called frequent itemsets. We can define 'item set' as a non-empty set of items (each context value is considered as an item in the mobile phone dataset). The cardinality of an item set ranges from one to any positive number, e.g., is greater than zero. Each transaction record in the dataset contains an item set of size n, i.e., if a transaction record contains three different items (I_1, I_2, I_3), then the size of the item set is 3. An item set that can be found frequently in a dataset is typically called a frequent itemset, which identified the minimum support threshold. For instance, if a threshold is set to identify the frequent or infrequent item sets, then the item sets that are observed below this minimum support threshold are called infrequent itemsets. On the other hand, the item sets that are observed with a higher value of this minimum support threshold are called frequent itemsets. Both frequent and infrequent itemsets are subsets of a superset and (ii) the second problem is to generate association rules from those frequent itemsets with another constraint of minimal confidence. Association rules are discovered from only the frequent itemsets that are discovered using the minimum support threshold discussed above. Thus, the discovery of a frequent itemset affects the number of discovered association rules. To determine whether an item set is frequent and infrequent, a minimum support threshold must be preset by the user. Otherwise, it is typically not possible to discover neither the frequent itemsets from a dataset nor their corresponding association rules.

Although association rule mining was introduced to extract associations from market basket data [77], association rules are employed today in many other domains such as data analysis, recommender systems, intrusion detection, and web usages mining etc. In the area of mining mobile phone data, recently, a number of researchers [25, 26, 37] also use association rules for various purposes. Many association rule mining algorithms have been proposed in the data mining literature, such as logic based [78], frequent pattern based [75, 79, 80], tree-based [81] etc. In the following, we mainly focus on some classic and popular association rule mining algorithms, such as AIS [77], Apriori [75], Apriori-based such as Apriori-TID and Apriori-Hybrid [75], FP-Tree [81], and RARM [82] algorithms.

3.3.1.1 AIS Algorithm

The AIS algorithm, proposed by Agrawal et al. [77], is the first algorithm designed for association rule mining. The main focus of this algorithm is to improve the quality of the datasets with the necessary functionalities for findings of associations or relations within this data and to process decision support queries using the discovered associations. In this algorithm, the consequent of the discovered association rules contains only one item, however, the antecedent may contain several items or contexts. An example of such association rule is like $A_1 \cap A_2 \Rightarrow C$, where A_1, A_2 are the items in antecedent and C (one item) represents the consequences of that rule.

In AIS algorithm [77], the frequent itemsets (each context value is considered as an item in the mobile phone dataset) were generated by scanning the datasets several times. A frequent item set satisfies the minimal support. This algorithm works based on several iterations or passes over the dataset. While processing, during the first pass over the dataset, the support count of each context value (item) was accumulated. The context that is infrequent gets eliminated from the list of items. An item is considered infrequent if it has a support count less than its predefined minimum support value according to the preference of the individual user. In such a way, candidate 1-item sets are generated from the dataset. After that, candidate 2-item sets are generated. To do this, this algorithm extends the generated frequent 1-item sets with the remaining other contexts available in the dataset during the second pass over the dataset. After that, similar to the first pass, the infrequent item set is eliminated in the second pass. For this, the algorithm again counts the support value of the generated candidate 2-item sets and checked with the same minimum support threshold that is preferred. The item sets whose support count do not satisfy this predefined threshold are also considered as infrequent item set. Similarly, based on the remaining other contexts in a record of the dataset, the $(n + 1)$ candidate itemsets are generated by extending the frequent n-item sets. The generation of all these candidate item sets and the corresponding frequent itemsets (identified by checking with the minimum support preference) identifying process iterate until any one of them becomes empty. Finally, this algorithm generates association rules based on the frequent itemsets that are identified in different iteration over the dataset.

To make AIS algorithm [77] more efficient, an estimation method was introduced to prune those generated item sets (combination of contexts) that cannot become frequent according to its support value. It not only prunes the unnecessary item sets candidates but also helps to avoid the corresponding unnecessary effort of counting those item sets. Besides such candidate generation, the memory management in the AIS algorithm is another issue, as all the candidate itemsets and frequent itemsets are assumed to be stored in the main memory. This is important when memory is not enough to store the huge amount of generated candidates. To resolve this issue memory management is also proposed for AIS; (i) to delete the generated candidate itemsets that have never been extended, (ii) to delete the generated candidate itemsets containing the maximal number of items and their siblings, and store this the parent item sets in the disk as a seed for the next pass, which is described with examples in [77].

The main drawback of the AIS algorithm is too many candidate item sets generation and consequently produce a huge number of redundant associations or rules. As a result, it not only produces several useless rules but also requires more space or memory related to such unnecessary generation and wastes much effort that turned out to be useless. At the same time, this algorithm needs too many passes over the whole dataset, which makes the AIS algorithm inefficient for mining mobile phone data to build context-aware real-life applications for mobile phone users.

3.3.1.2   Apriori Algorithm

Apriori proposed by Agrawal in [75] is the most popular algorithm in the area of mining association rules. According to [83], it is a great improvement in the history of association rule mining. The AIS algorithm [77] described above is just a straightforward association generation approach that requires many passes over the dataset, generating many candidate item sets while most of them turn out to be useless. Comparing with the AIS algorithm, Apriori is more efficient during the candidate generation process for two reasons. The first one is Apriori employs a new method deferring from the AIS algorithm, for generating the candidate itemsets (a set of context values), and the second one is it also introduces a new pruning technique for eliminating the infrequent candidates.

To generate all the candidate itemsets (a set of contexts) from a given dataset, there are two processes in Apriori algorithm [75]. Firstly, this algorithm generates the candidate itemsets using the available items in the dataset. After generating these candidates, the support value of the corresponding generated item sets is counted by scanning the dataset. While processing, during the first scanning over the dataset, the support count of each context value (item) is calculated to identify the frequent item set. A frequent itemset satisfies the minimal support preferred by an individual user. This algorithm performs a pruning operation and prunes the infrequent item sets to reduce the burden of further processing. An item is considered infrequent if it has a support count less than its predefined minimum support value set by the preference of an individual user. On the other hand, the item sets that satisfy this threshold are considered as frequent item sets, which are checked in each iteration of the algorithm and generates only those candidate item sets that include the same specified number of items, such as 1-context set, 2-context set, etc. In such a way, the candidate n-item sets are generated after the $(n-1)$th passes. To do this, this algorithm performs the joining operation with only the frequent $(n-1)$-item sets by counting their corresponding support value over the dataset. To generate these candidate item sets, the Apriori algorithm follows its property, which is defined as "every sub $(n-1)$-item sets of the frequent n-item sets must be frequent" [75]. As such, if any sub $(n-1)$-item sets are not in the list of frequent $(n-1)$-item sets, then the n-item sets candidate is pruned. According to this condition, all the candidate n-item sets are pruned by checking their sub $(n-1)$-item sets during the process as there is no possibility to be frequent according to the Apriori property [75].

In summary, Apriori algorithm [75] avoids the effort wastage of counting the candidate itemsets (a set of contexts) that are already known to be infrequent (not satisfy the support threshold), during the process of identifying frequent itemsets. This algorithm not only generates the candidate itemsets by joining among the frequent itemsets (satisfy the support threshold preferred by an individual) level-wisely but also prunes the candidates according to the Apriori property that is mentioned above. As a result, the number of remaining candidate item sets becomes much smaller, which are ready for further processing. It dramatically reduces the computation, I/O cost and memory requirement comparing with the AIS algorithm [77]. However, the Apriori still has two major drawbacks, of which one has been inherited from the AIS approach. The inherited drawback is that it still has to scan the entire dataset multiple times as it builds the list of frequent itemsets, which eventually produces a huge number of redundant rules. The second drawback is that the candidate generation process is time and memory-consuming and complex that is not effective for mining mobile phone data to build context-aware real-life applications for mobile phone users.

3.3.1.3 Apriori-Based Algorithms

Based on the Apriori algorithm [75], several new association rule mining algorithms were designed with some modifications of this algorithm. For example, Apriori-TID and Apriori-Hybrid [75] are the modifications of the Apriori algorithm.

These algorithms are based on the Apriori algorithm and try to improve the efficiency in terms of execution time by making some modifications. These algorithms try to reduce the number of passes over the dataset and to reduce the size of the dataset to be scanned in every pass for generating the candidate itemsets. Also, these algorithms try to prune the generated candidates by using different techniques.

Apriori-TID [75] extends the original Apriori algorithm by removing the need for multiple scanning of the datasets. This algorithm sets a counter during the first pass through the dataset. This counter is then used later to determine the frequent itemsets. As a result, the original dataset is not needed to counter this. On the other hand, Apriori-Hybrid [75] is based on the idea that is not necessary to use a similar process for each pass over the dataset for generating candidates. This approach combines the advantage of using the Apriori algorithm in the early passes and later this algorithm uses the Apriori-TID algorithm. There is a problem with this approach is the cost of switching between Apriori and Apriori-TID algorithms. Another algorithm Predictive Apriori proposed by Scheffer [84] generates rules by predicting predictive accuracy combining form support and confidence. So sometimes it produced the rules with large support but low confidence and got unexpected results. These algorithms could give better results in terms of computational time in some cases. However, the problem of redundancy still exists in these modified algorithms.

3.3.1.4   FP-Tree Algorithm

Frequent Pattern Tree (known as FP-Tree) is another association rule mining algorithm, which was introduced by Han et al in [81], to mine frequent patterns like Apriori. FP-Tree is another milestone in the development of mining association rules in terms of execution time. This algorithm needs no candidate generation process like AIS and Apriori algorithm. It can generate frequent itemsets with only two passes over the dataset. The frequent patterns generation process includes two sub-processes: (i) it first constructs the frequent pattern tree, and (ii) then generates the frequent patterns from the tree. By avoiding the candidate generation process and taking less scanning over the dataset, FP-Tree becomes an order of magnitude faster algorithm than the AIS [77] and Apriori [75] algorithm for generating frequent patterns from a given dataset [83].

Three reasons make the FP-Tree algorithm more efficient [83]. First, this algorithm generates a frequent pattern tree, which is a compressed representation of a given dataset. While constructing the tree, only frequent items measured by counting the support value are used. A frequent itemset satisfies the minimal support preferred by an individual user. The other irrelevant information is pruned. This algorithm also does ordering the items (contexts) according to their support values [81]. Secondly, this algorithm only scans the dataset twice. The patterns that satisfy the user-specified minimum threshold are generated by constructing the conditional FP-Tree. To do this, this algorithm uses the concept of the suffix of the patterns. For instance, the conditional FP-Tree contains only the patterns with the specified suffix of the patterns, which reduces the computation cost dramatically. Thirdly, the frequent pattern tree uses a divide and conquer method. It considerably reduces the size of the subsequent conditional tree. This algorithm generates the longer frequent patterns by extending the shorter patterns, i.e., adding a suffix to the shorter frequent patterns [83].

Although the FP-tree does not generate candidates like Apriori, it produces similar outputs for the same dataset. As a result, the problem of producing redundant association rules still exists. Thus, it will not be effective for mining mobile phone data to build context-aware real-life applications for mobile phone users because of its redundant association generation [83].

3.3.1.5   RARM Algorithm

Rapid Association Rule Mining (RARM) is an approach proposed by Das et al. [82] that also uses the tree structure to represent the dataset. This approach also does not utilize a candidate generation process. RARM algorithm uses the SOTrieIT (Support Ordered Trie Itemset) structure for generating the candidate itemsets such as 1-item sets and 2-item sets. It can quickly generate such item sets without generating the candidates and scanning the dataset as well. Similar to the FP-Tree [81] that is discussed above, every node of the SOTrieIT contains one item and the corresponding support value. This algorithm uses the SOTrieIT tree for generating the candidate item-sets [82].

The main focus of this algorithm is faster processing than the existing algorithms. According to [82] RARM is up to 100 times faster than Apriori [75]. However, the problem of producing redundant association rules still exists. Thus, it will not be effective for mining mobile phone data to build context-aware real-life applications for mobile phone users because of its redundant association generation [83].

3.3.1.6   Association Rule Mining Summary

The association rule learning algorithm finds association rules from a dataset that satisfy predefined minimum support and confidence constraints [75]. In the literature on data mining, several association rule learning algorithms have been suggested, such as logic-based [78], frequent pattern based [75, 79, 80], tree-based [81] etc. As it has its parameter support and confidence, the association rule learning technique is well established in terms of rule efficiency, e.g., accuracy and flexibility [85]. A number of researchers [25, 26, 37] have used association rule learning technique (e.g., Apriori)[75] to mine rules capturing mobile phone users' behavior. However, when it comes to discovering users' behavioral rules, association rule learning has some limitations. The disadvantages of association rules for discovering the behavioral rules of individual mobile phone users when taking into account multi-dimensional contexts are summarized below.

- *Lacking in Understanding the Impact of Contexts:* Different contexts in mobile phone data, such as temporal, spatial, or social background, can have varying effects or influence on individual mobile phone users' behavioral rules. Incoming phone calls from a significant person, such as a mother, are often answered by a person, even if she is in a meeting since her family comes first. In this case, the importance of individuals' social relationships (*social relationship* → *mother*) in making behavioral decisions is greater than other related contexts such as time, weekday or holiday, place, accompanied with, and so on. However, when discovering rules based on multi-dimensional contexts, the standard association rule learning methodology implicitly assumes that all of the contexts in the datasets have the same nature and/or effect.

- *Redundancy:* If a user preference is defined as a minimum support value and minimum confidence value, an association rule learning technique, such as Apriori, discovers all the contextual associations in a given dataset. As a consequence, the association rule learning methodology generates a large number of redundant rules because it does not consider the usefulness of the associations when creating them. For instance, it produces up to 83% redundant rules for a given dataset that makes the rule-set unnecessarily large [86]. Therefore, it is very difficult to determine the most interesting ones among the huge amount of rules generated. As a result, it makes the rule-based decision-making process ineffective and more complex, which is not effective to build a context-aware intelligent system [87].
- *Computational Complexity and High Training Time:* The association rule learning method necessitates a significant amount of preparation time to generate rules. For example, when the association rule learning algorithm is used to discover user behavioral rules, the authors find a long-running period spanning many hours in an experimental study in the cell phone domain [37]. The key explanation for the long training period is that traditional association methods compute all possible correlations between contexts and are unable to filter out the useful rules for decision-making. As a consequence, generating patterns that aren't required adds to the computational complexity and training time.

In summary, traditional association rule learning techniques may not be appropriate to generate users' behavioral rules in multi-dimensional environments, to create intelligent context-aware systems, when taking into account the effect of contexts, the redundancy issue while producing rules, and computational complexity.

3.3.2   Classification Rules

Another method for extracting user behavioral rules from datasets is classification. Classification is another tool for discovering rules in the field of data mining, where A represents contextual information and C represents the corresponding activity class. In general, classification is classified as a learning method for mapping (classifying) a data instance into the dataset's predefined class labels. According to [88], data classification is a two-step process; (i) is the learning stage, in which a classification model is built from a given dataset; the training set is the data from which a classification feature or model is learned, and (ii) second one is a classification step, in which the model is used to evaluate or predict class labels for previously unknown data; the testing set is the data set used to test the learned model's or function's classification ability.

Classifier efficiency is normally determined by accuracy, which is the percentage of correct predictions over the total number of predictions made for a given test dataset, according to [88]. Many other metrics, such as sensitivity, error rate, specificity, precision, recall, and f-measure, are also used to understand the various aspects of the generated model. In the data mining literature, several classification algorithms with rule generation capabilities have been proposed. In the following, we mainly focus on some basic and popular approaches such as ZeroR [89], OneR [90], Decision Tree [76], PART [91], and RIDOR [89].

3.3.2.1 Zero-R

Among the classification techniques used in the data mining field, Zero-R is the most basic [89]. This algorithm only considers the objective and disregards all predictors. The majority group is predicted by the Zero-R classifier (class). For example, a Zero-R model for a sample phone call dataset may be *"behavior → reject"*. Although Zero-R has no predictability capacity, it can be used to establish a baseline output for other classification methods [89].

3.3.2.2 One-R

Holte et al. [90] proposed One-R, which is a simple and inexpensive classification method. One-R, short for "One Rule," is a straightforward but precise classification algorithm for generating the predicting rule. In this method, a one-level decision tree is built from the training records, and the rules are extracted from that tree, which is connected to frequently occurring classes in the dataset. Humans can easily interpret and comprehend the rules produced by the One-R approach. This algorithm creates a frequency table for each predictor against the target to generate a rule for it. It then produces one rule for each predictor in the data and chooses the rule with the smallest total error as its "one rule". If a user is in a meeting, for example, the phone call action is rejected. One-R has been shown to generate rules that are only marginally less reliable than state-of-the-art classification algorithms while still being easy to understand by humans [90].

3.3.2.3 RIDOR

A direct classification tool is the Ripple Down Rule learner (RIDOR) [89]. The default rule and the additional rules of that default rule are created by this algorithm. The algorithm produces a default rule by evaluating the dataset first, according to this principle. Following that, it generates a set of additional rules. To do this, the algorithm measures the error rate and selects the rules with the lowest error rate. These created additional rules are used in addition to the default rule to predict the unseen classes for a given condition.

3.3.2.4 Decision Tree

Decision trees [92] is a well-known and widely discussed technique for classification and prediction. A decision tree is a graph that illustrates the possible outcome of a decision using a branching process. Each branch of a decision tree represents a choice among a set of options related to the context of attribute values, and each leaf node represents a classification or decision for that choice. The decision tree algorithm aims to derive classification rules from instance learning. We'll go through some simple decision tree algorithms in this section. Various algorithms exist in the implementation of speed, scalability, performance intelligibility, classification accuracy, and other factors, each with its own set of advantages.

J. R. Quinlan suggested ID3 as the central algorithm for creating decision trees [92]. The ID3 algorithm builds a decision tree by using a top-down approach in which each attribute or context is tested at each node using a greedy search through the specified training dataset. It determines the entropy and information gain, which is a statistical property used to determine which attribute to measure at each node in the tree [92]. The degree to which a given attribute distinguishes training examples according to their target classification is measured by information gain. For both missing values and continuous-valued attributes, the ID3 algorithm does not produce sufficient results. The values for a continuous attribute should be mapped to some discrete representation to improve ID3 efficiency.

Quinlan proposes a modified algorithm, the C4.5 algorithm, based on the ID3 algorithm [76]. C4.5 uses the principle of knowledge benefit to construct decision trees from a training dataset in a similar manner to ID3. For splitting the dataset into subsets, the C4.5 decision tree algorithm uses gain ratio as the test attribute selection criterion, and each time selects the attribute with the highest knowledge gain ratio as the test attribute for a given set. C4.5 is a statistical classifier that can deal with both numeric and missing value attributes, is robust in the presence of noise, and can build trees with large branches and scales.

The CART algorithm tends to simplify and increase the performance of the decision tree [93]. CART will deal with both order and disorder data in addition to multi-state numerical data. It chooses the test attribute based on the Gini coefficient and creates a binary tree with a straightforward structure. Following that, the SLIQ [94] and SPRINT [95] algorithms are proposed solely to improve scalability and parallelism.

3.3.2.5 Hybrid Classification

In [91], PART was proposed as a hybrid classification algorithm. This algorithm generates rules using a rule induction method in addition to the decision tree. Instead of using these two algorithms in two steps, this algorithm combines them all into one. Even though this algorithm creates a decision tree, it does not create a complete decision tree. A partial decision tree is built using a divide and conquer method in

PART. A rule induction procedure is used to produce the candidate rules. After that, this algorithm employs a pruning process to filter the intended rules. DTNB, proposed by Sheng et al. [96], is another hybrid classification technique for creating classifications, similar to the PART algorithm. It employs a decision table as well as a Naive Bayes classifier. The generated rules can be used to predict previously unknown classes in a given situation.

3.3.2.6   Classification Rule Mining Summary

One of the most common rule-based classification algorithms is decision tree, which has several advantages, including being easier to interpret, being able to handle high-dimensional data, being simple and quick, being accurate, and being able to generate human-understandable classification rules [97, 98]. A number of authors [36, 99–101], in particular, have used the decision tree classification method to find rules capturing cell phone users' actions for different purposes. However, to model users' actions, classification rules have some limitations. The disadvantages of rule-based classification strategies for discovering the behavioral rules of individual cell phone use are summarized below.

- *Low-Reliability:* In general, reliability refers to the consistency of being dependable or continuously performing well. If the pattern or rule describes a relationship that happens in a high percentage of relevant situations, it is said to be accurate. A classification rule will be reliable if it provides high prediction accuracy, and an association rule will be reliable if it has high confidence correlated with the accuracy, according to Geng et al. [102]. However, in many situations, the classification rules discovered by traditional rule-based classification methods, such as decision trees, have poor reliability [25, 103]. A classification rule does not guarantee high accuracy in forecasts, according to Freitas et al. [85]. The explanation for this is that it may have an over-fitting problem and inductive bias, both of which reduce the accuracy of a machine learning-based model's prediction.
- *Lacking in Flexibility:* Traditional rule-based classification techniques, such as decision trees, do not allow users to set their preferences, and as a result, they make rigid decisions for each test case [76]. However, when considering real-world scenarios, static decisions in modeling user actions can not be meaningful. The explanation for this is that people's preferences aren't always consistent; they can differ from one user to another [104]. For example, one person will wish for the phone call agent to reject incoming calls if she has not answered them more than 80% of the time in the past. This preference could be 95% of the time for another person, depending on her preferences.
- *Lacking in Generalization:* Typically, generality refers to the extent to which a pattern or rule is systematic, i.e., the percentage of all applicable records in the dataset that match the pattern. According to Geng et al. [102], a pattern is more useful and interesting if it characterizes more details in the related

dataset. When creating classification rules, traditional classification strategies take data-driven generalization into account. Aside from that, users' behavior-based generalization might be of concern. Users' actions, for example, can be consistent in a variety of situations, with just a few exceptions [105]. As a result, behavior-oriented generalization may provide more accurate results for modeling users' use patterns. The generalization process not only simplifies the resulting machine learning-based model but also reduces overfitting and increases prediction accuracy.

Neither association rule mining (e.g., Apriori) [75] nor classification rule mining (e.g., Decision tree) [76] are ideal for discovering behavioral rules of cell phone users based on multi-dimensional contexts. As a result, in this book we suggest a behavioral decision tree-based approach that generates not only useful general rules for capturing individual actions at a given level of confidence with a small number of contexts but also rules that articulate unique exceptions to the general rules when more context-dimensions are considered.

3.4   Incremental Learning and Updating

Mobile phone log data is not static; it is gradually applied day by day based on an individual's current (on-going) mobile phone habits. Since people's behavior changes over time, the more recent trends are more likely to be important and relevant for predicting people's potential behavior in specific situations than older ones [106, 107]. As a result, updating and complex management of discovered rules based on individuals' recent behavioral trends (e.g., recency) becomes a challenge, as changes can not only invalidate certain current rules but also make other rules relevant.

Several incremental rule mining techniques have been proposed for mining rules in a complex database in the field of data mining. To get a fully updated set of rules, these techniques use existing rules and the incremental portion of the dataset. For example, Cheung et al. [108] proposed the FUP algorithm, which is the first incremental updating technique for preserving association rules as new data is added into the database. The FUP algorithm is used to discover new frequent itemsets in a complex database and is based on the Apriori [75] algorithm.

By deleting earlier itemsets that are either considered to be still frequent or deemed infrequent only by testing the incremental section, FUP tries to extract the value from the previously discovered rules to produce a relatively small candidate set to be tested against the original data set. Cheung, et al. suggested a new algorithm FUP2, which is an expansion of the FUP algorithm, in [109]. When new transactions are added to a database, the FUP updates the association rules, while the FUP2 extracts the rules from the final data collection, taking into account both the removed and newly added parts. If the data set is only modified by insertions, the FUP2 algorithm would behave similarly to FUP.

Xu et al. [110] suggest another incremental association rule mining algorithm. They suggest an IFP-tree technique, which is an extension of the FP-tree technique [81]. When a data set is incremented, it uses the current FP-tree to construct the IFP-tree. It updates the support of the related nodes of the tree for each transaction of the incremental part. If, on the other hand, a new node is formed, it is held as a new branch of the root node. The frequent itemsets are discovered after the tree has been built. It does so by contrasting both old and new trees. When a new node in the updated tree is discovered, frequent itemsets are created. Since each new node in the tree forms its branch, only the new branches can result in the generation of a new frequent object. However, as the number of dimensions and transactions grows, its efficiency suffers. Thomas et al. propose an algorithm based on the idea of Negative Border that preserves both frequent and border itemsets [111]. The algorithm updates to support counts of all frequent itemsets and border itemsets as data is added to or removed from the original database. However, to minimize scanning times of an initial database, a large number of border itemsets must be stored in memory.

A few algorithms are proposed based on the three-way decision, which is an extension of the widely used binary-decision model with an optional third alternative [112, 113]. The concepts of approval, rejection and no engagement suggested by Yao [114] are used to build a three-way decision theory. All itemsets are divided into three regions using these methods, namely the positive, the boundary, and the negative region. Positive itemsets are already popular, and they consider them without reservation. Itemsets in the boundary region are uncommon, but they could become more common shortly after data increment. Even after data increment, itemsets in the negative region will become less common, and they will be abandoned. As a result, all that is required to keep the frequently used itemsets up to date is to review those in the boundary zone. The runtime is saved because the negative region containing the majority portion is never computed.

Amornchewin et al. suggest a probability-based incremental association rule discovery algorithm in [115]. To avoid reprocessing entire dynamic databases, this algorithm uses the Bernoulli trials principle and uses previously mined information. When only new data is inserted into a dynamic database, the algorithm can efficiently maintain association rules. Thusaranon et al. [116] propose a new probability-based incremental association rule discovery algorithm that is a development of Amornchewin and Kreesuradej's [115] algorithm. They expand the algorithm's ability to maintain association rules of a complex database in the case of record insertion and deletion at the same time in this algorithm.

The above incremental mining techniques primarily consider the overall mining process's faster processing, e.g., performance. Instead of processing the combined dataset that includes the initial dataset and the incremental portion, these techniques minimize scanning on the provided datasets by mining the incremental part separately. As a result, the overall mining method with conventional updating techniques affects the amount of time it takes to find a full set of revised rules. However, the freshness of rules, such as rules based on recent trends, is important in modeling users' behavior, and this has not been taken into account in these techniques. The explanation for this is that in the real world, user behavior is not static and can change over time. To effectively model smartphone users' behavior in relevant multi-dimensional contexts, updation in terms of freshness in users' behavior while generating rules is needed.

Several researchers use recent cell phone log data behavioral patterns to forecast future actions rather than patterns extracted from the entire historical logs to generate rules based on an individual's current behavior. They do, however, use a static period of recent historical data, which may be insufficient for determining users' recent behavioral rules. For example, Lee et al. [117] have used recent call list data to study cell phone users' calling patterns and create a call recommendation algorithm for an adaptive speed-call list. To achieve their target, they extract call logs from the previous 3 months. Barzaiq et al. [118] suggest an approach that analyzes cell phone historical data over a 2-year cycle to forecast outgoing calls and observe relatively additional computational load that appears to be unnecessary. Phithakkitnukoon et al. [119], conduct their research on reality mining datasets collected over 9 months and find that only a recent portion of contact history is more important. Phithakkitnukoon et al. [65] present a model for forecasting phone calls for the next 24 h based on the users' previous contact history in a separate paper. They demonstrated that the recent trend of the user's calling pattern is more important than the older one and has a higher correlation to the future pattern than the pattern generated from all historical data in their approach. As a result, to improve prediction accuracy, the most recent 60 days of call records in the call logs are considered to be the potential observed call activities [65]. However, since users' actions are not consistent in the real world and may differ from user to user over time, such a static period consideration may not be appropriate to represent one's current behavior.

Apart from these methods, several authors [120, 121] deal with the issue of handling personal information, such as individual's contact lists in their mobile phone, and more precisely, the task of searching for the desired contact number when making an outgoing call. According to Bergman et al. [120], a large number of contacts in cell phones are never used, even though contact lists grow larger. According to their findings, 47% of the users' contacts had not been used in over 6 months or had never been used at all. Stefanis et al. [121] used a window-based model for handling and searching personal information on mobile phones to predict future actions. They demonstrated in their experiment that the training window for predicting an individual's cell phone use behavior should be long enough to provide enough data. A training window of more than 2 weeks, on the other hand, would be unable to capture the dynamic changes in phone call behavior patterns. Furthermore, a training window of fewer than 7 days will be insufficient to capture behavioral changes through all days of the week, including changes in social circumstances on weekends. In conclusion, standard updating techniques discussed above may not be appropriate for producing a full collection of users' behavioral rules in multi-dimensional contexts, to develop intelligent context-aware systems, to provide relevant services to end smartphone users, when taking into account the freshness in rules representing users' current actions and their dynamic updation.

3.5 Identifying the Scope of Research

Mobile phones have become an inseparable part of our lives. As mentioned in this chapter, not all users' cell phone use behaviors are the same and can vary from one user to another based on contextual details. In a context-aware pervasive computing environment, studies have demonstrated that mining contextual behavior rules of individual cell phone users will intelligently assist them in various context-aware personalized systems such as smart call interruption management system, smart call reminder system, mobile notification management system, context-aware mobile app recommender systems, and various predictive services.

Users do not want to use apps that require individual cell phone users to identify and maintain their behavioral rules manually rather than automatically discovered, according to studies. Users may not have the time, inclination, experience, or interest to manually maintain rules [122]. Mining behavioral rules of individual cell phone users based on relevant multi-dimensional contexts, according to individual preferences, is a key prerequisite for developing such smart applications.

The state-of-the-art in the field of mobile data analytics was addressed in this chapter. Based on this, we have summarized the scope of research below, which are taken into account throughout this book:

(i) Contextual data pre-processing and feature selection are the primary parts of an effective context-aware system. In Chap. 4, the basic feature selection and extraction methods for efficient processing have been provided. We also present several contextual datasets that can be utilized to build a machine learning based context-aware model for corresponding mobile applications and services in this chapter. As the real-world data may contain noisy and inconsistency instances, the pre-processing steps have also been analyzed to clean and remove noises from raw data in this chapter.

(ii) Context discretization, e.g., an effective time-series modeling considering mobile user behavioral activities is still lacking for building an intelligent context-aware system. In Chap. 5, we present a behavior-oriented time segmentation technique capturing user behavioral patterns to produce temporal behavioral rules. Using time-series cell phone data, this method dynamically considers not only the temporal coverage of the week but also the number of incidences of various behaviors to produce related behavioral time segments over the week.

(iii) Existing studies have focused on mainly association rule mining techniques (e.g., Apriori) or classification rule mining techniques (e.g., Decision tree) for discovering user behavioral rules utilizing mobile phone data. However, there is still a lack of discovering the useful behavioral rules of individual mobile phone users based on multi-dimensional contexts. In Chap. 6, we present a tree based approach to model an individual's mobile phone usage behavior utilizing their mobile phone data. This approach produces not only the general rules that capture an individual's behavior at a particular level of confidence with a minimal number of contexts but also produces rules that express specific exceptions to the general rules when more context dimensions are taken into account.

(iv) Another key limitation of the existing literature is that the previous studies have not considered the recency-based rules of individual mobile phone users and updating the existing rules based on the recent behavioral patterns of individuals. In other words, there is a lack of understanding about rules that reflect an individual's recent behavioral patterns ignoring the user's past behavioral patterns to get a complete set of updated rules. In Chap. 7, we present an approach for recency-based updating the rules and their dynamic management. This updated rules-set not only contains all the useful rules of an individual mobile phone user for the whole log period but also expresses recent behavioral patterns that will help model mobile phone usage behavior of individuals to provide personalized services for the end mobile phone users in a context-aware pervasive computing environment.

(v) A rule-based expert system modeling is typically considered one of the key AI techniques that can be used to make intelligent decisions and more powerful applications. In Chap. 8, we discuss mobile expert system as a knowledge or rule-based modeling, where a set of context-aware rules are extracted from mobile data discussed in earlier chapters. Usually, the purpose of the expert system is to take information from a human expert and turn this into a number of hardcoded rules for the input data to be implemented. In this chapter, we focus on the generated rules based on machine learning techniques, rather than the hardcoded rules, as we take into account the dynamism in the context-aware rules.

(vi) Deep learning is part of a broader family of machine learning methods based on artificial neural networks with representation learning. Although, the rule-based machine learning methods performed well, deep learning can be used, when a large amount of data is available. In Chap. 9, we discuss the importance of deep learning and a context-aware deep learning model for mobile phone users.

(vii) Finally, in Chap. 10, we highlight the most important and vital issues, ranging from contextual data collection to decision-making, that has been thoroughly explored in this book. In terms of new researchers' perspectives, future advances in industries, and smart solutions in the context-aware technology industry, prospective research works, and challenges in the field of context-aware computing have been addressed.

Overall, this book presents a variety of techniques and research scope in the field of context-aware machine learning and data analytics, which can be used for building a variety of real-world applications. The prominent application fields are personalized assistance services, recommendation systems, human-centric computing, adaptive and intelligent systems, IoT services, smart cities, mobile privacy and security systems, and many others, where applications dynamism based on contextual data is needed.

3.6 Conclusion

The effect of multi-dimensional contexts on smartphone data, both in terms of context-aware rule learning and the dataset itself, was addressed in this book. In this chapter, we mainly aimed to address the current state of mobile data analytics and associated rule learning techniques. We have also explored about how multi-dimensional contexts including temporal, spatial, and social contexts can influence such technology. We have summarized applicable research for each popular technique to help others in the context-aware rule learning community. In terms of time-series modeling, rule discovery based on multi-dimensional contexts, and updating rules over time according to individual preferences, we have addressed various issues regarding context-aware rule learning. In terms of current research, there has been a lot of emphasis on conventional context-aware systems and techniques, with less work on machine learning rule-based context-aware systems for successful decision making in a specific domain.

Overall, we reviewed previous research and addressed a discussion of challenges and potential directions for learning context-aware rules from smartphone data in this chapter. The domain-specific context-aware rules can be used to create a variety of context-aware models that intelligently assist end-users in their daily activities. At the end of this chapter, we have summarized the scope of research, which are taken into account throughout this book. We also assume that this study can be used as a reference guide in the relevant application areas including mobile applications, smart systems and security, etc. for both academia and industry.

References

1. Dourish, P. (2004). What we talk about when we talk about context. *Personal and Ubiquitous Computing, 8*(1), 19–30.
2. Schilit, B. N., & Theimer, M. M. (1994). Disseminating active map information to mobile hosts. *IEEE network, 8*(5), 22–32.
3. Brown, P. J., Bovey, J. D., & Chen, X. (1997). Context-aware applications: From the laboratory to the marketplace. *IEEE personal communications, 4*(5), 58–64.
4. Ryan, N. S., Pascoe, J., & Morse, D. R. (1998). Enhanced reality fieldwork: The context-aware archaeological assistant. In *Computer applications in archaeology*. Tempus Reparatum.
5. Brown, P. J. (1995). The stick-e document: A framework for creating context-aware applications. *Electronic Publishing-Chichester-, 8*, 259–272.
6. Franklin, D., & Flaschbart, J. (1998, March). All gadget and no representation makes Jack a dull environment. In *Proceedings of the AAAI 1998 Spring Symposium on Intelligent Environments* (pp. 155–160).
7. Ward, A., Jones, A., & Hopper, A. (1997). A new location technique for the active office. *IEEE Personal communications, 4*(5), 42–47.
8. Hull, R., Neaves, P., & Bedford-Roberts, J. (1997, October). Towards situated computing. In *Digest of papers. First international symposium on wearable computers* (pp. 146–153). IEEE.
9. Rodden, T., Cheverst, K., Davies, K., & Dix, A. (1998, May). Exploiting context in HCI design for mobile systems. In *Workshop on human computer interaction with mobile devices* (Vol. 12).

10. Schilit, B., Adams, N., & Want, R. (1994, December). Context-aware computing applications. In *1994 first workshop on mobile computing systems and applications* (pp. 85–90). IEEE.
11. Dey, A. K. (2001). Understanding and using context. *Personal and Ubiquitous Computing, 5*(1), 4–7.
12. Chang, Y. J., & Tang, J. C. (2015, August). Investigating mobile users' ringer mode usage and attentiveness and responsiveness to communication. In *Proceedings of the 17th International Conference on Human-Computer Interaction with Mobile Devices and Services* (pp. 6–15).
13. Halvey, M., Keane, M. T., & Smyth, B. (2006, April). Time based patterns in mobile-internet surfing. In *Proceedings of the SIGCHI Conference on Human Factors in Computing Systems* (pp. 31–34).
14. Gandhi, S., Oates, T., Boedihardjo, A., Chen, C., Lin, J., Senin, P. & Wang, X. (2015, October). A generative model for time series discretization based on multiple normal distributions. In *Proceedings of the 8th Workshop on Ph. D. Workshop in Information and Knowledge Management* (pp. 19–25).
15. Farrahi, K., & Gatica-Perez, D. (2014). A probabilistic approach to mining mobile phone data sequences. *Personal and Ubiquitous Computing, 18*(1), 223–238.
16. Zhang, G., Liu, X., & Yang, Y. (2014). Time-series pattern based effective noise generation for privacy protection on cloud. *IEEE Transactions on Computers, 64*(5), 1456–1469.
17. Song, Y., Ma, H., Wang, H., & Wang, K. (2013, May). Exploring and exploiting user search behavior on mobile and tablet devices to improve search relevance. In *Proceedings of the 22nd International Conference on World Wide Web* (pp. 1201–1212).
18. Rawassizadeh, R., Momeni, E., Dobbins, C., Gharibshah, J., & Pazzani, M. (2016). Scalable daily human behavioral pattern mining from multivariate temporal data. *IEEE Transactions on Knowledge and Data Engineering, 28*(11), 3098–3112.
19. Mukherji, A., Srinivasan, V., & Welbourne, E. (2014, September). Adding intelligence to your mobile device via on-device sequential pattern mining. In *Proceedings of the 2014 ACM International Joint Conference on Pervasive and Ubiquitous Computing: Adjunct Publication* (pp. 1005–1014).
20. Bayir, M. A., Demirbas, M., & Cosar, A. (2011). A web-based personalized mobility service for smartphone applications. *The Computer Journal, 54*(5), 800–814.
21. Paireekreng, W., Rapeepisarn, K., & Wong, K. W. (2009). Time-based personalised mobile game downloading. In *Transactions on edutainment II* (pp. 59–69). Berlin, Heidelberg: Springer.
22. Jayarajah, K., Kauffman, R., & Misra, A. (2014, September). Exploring variety seeking behavior in mobile users. In *Proceedings of the 2014 ACM International Joint Conference on Pervasive and Ubiquitous Computing: Adjunct Publication* (pp. 385–390).
23. Do, T. M. T., & Gatica-Perez, D. (2010, December). By their apps you shall understand them: Mining large-scale patterns of mobile phone usage. In *Proceedings of the 9th International Conference on Mobile and Ubiquitous Multimedia* (pp. 1–10).
24. Xu, Y., Lin, M., Lu, H., Cardone, G., Lane, N., Chen, Z.,& Choudhury, T. (2013, September). Preference, context and communities: A multi-faceted approach to predicting smartphone app usage patterns. In *Proceedings of the 2013 International Symposium on Wearable Computers* (pp. 69–76).
25. Mehrotra, A., Hendley, R., & Musolesi, M. (2016, September). PrefMiner: Mining user's preferences for intelligent mobile notification management. In *Proceedings of the 2016 ACM International Joint Conference on Pervasive and Ubiquitous Computing* (pp. 1223–1234).
26. Zhu, H., Chen, E., Xiong, H., Yu, K., Cao, H., & Tian, J. (2014). Mining mobile user preferences for personalized context-aware recommendation. *ACM Transactions on Intelligent Systems and Technology, 5*(4), 1–27.
27. Oulasvirta, A., Rattenbury, T., Ma, L., & Raita, E. (2012). Habits make smartphone use more pervasive. *Personal and Ubiquitous Computing, 16*(1), 105–114.

28. Yu, K., Zhang, B., Zhu, H., Cao, H., & Tian, J. (2012, May). Towards personalized context-aware recommendation by mining context logs through topic models. In *Pacific-Asia conference on knowledge discovery and data mining* (pp. 431–443). Berlin, Heidelberg: Springer.
29. Naboulsi, D., Stanica, R., & Fiore, M. (2014, April). Classifying call profiles in large-scale mobile traffic datasets. In *IEEE INFOCOM 2014-IEEE conference on computer communications* (pp. 1806–1814). IEEE.
30. Dashdorj, Z., Serafini, L., Antonelli, F., & Larcher, R. (2013, December). Semantic enrichment of mobile phone data records. In *Proceedings of the 12th International Conference on Mobile and Ubiquitous Multimedia* (pp. 1–10).
31. Shin, D., Lee, J. W., Yeon, J., & Lee, S. G. (2009, July). Context-aware recommendation by aggregating user context. In *2009 IEEE conference on commerce and enterprise computing* (pp. 423–430). IEEE.
32. Shin, C., Hong, J. H., & Dey, A. K. (2012, September). Understanding and prediction of mobile application usage for smart phones. In *Proceedings of the 2012 ACM Conference on Ubiquitous Computing* (pp. 173–182).
33. Farrahi, K., & Gatica-Perez, D. (2010). Probabilistic mining of socio-geographic routines from mobile phone data. *IEEE Journal of Selected Topics in Signal Processing, 4*(4), 746–755.
34. Ma, H., Cao, H., Yang, Q., Chen, E., & Tian, J. (2012, April). A habit mining approach for discovering similar mobile users. In *Proceedings of the 21st International Conference on World Wide Web* (pp. 231–240).
35. Cao, H., Bao, T., Yang, Q., Chen, E., & Tian, J. (2010, October). An effective approach for mining mobile user habits. In *Proceedings of the 19th ACM International Conference on Information and Knowledge Management* (pp. 1677–1680).
36. Zulkernain, S., Madiraju, P., Ahamed, S. I., & Stamm, K. (2010). A mobile intelligent interruption.
37. Srinivasan, V., Moghaddam, S., Mukherji, A., Rachuri, K. K., Xu, C., & Tapia, E. M. (2014, September). Mobileminer: Mining your frequent patterns on your phone. In *Proceedings of the 2014 ACM International Joint Conference on Pervasive and Ubiquitous Computing* (pp. 389–400).
38. Parate, A., Böhmer, M., Chu, D., Ganesan, D., & Marlin, B. M. (2013, September). Practical prediction and prefetch for faster access to applications on mobile phones. In *Proceedings of the 2013 ACM international joint conference on Pervasive and Ubiquitous Computing* (pp. 275–284).
39. Myllärniemi, V., Korjus, O., Raatikainen, M., Norja, T., & Männistö, T. (2014, November). Meeting scheduling across heterogeneous calendars and organizations utilizing mobile devices and cloud services. In *Proceedings of the 13th International Conference on Mobile and Ubiquitous Multimedia* (pp. 224–227).
40. Alexiadis, A., & Refanidis, I. (2009, April). Defining a task's temporal domain for intelligent calendar applications. In *IFIP international conference on artificial intelligence applications and innovations* (pp. 399–406). Boston, MA: Springer.
41. Sarker, I. H., Colman, A., Han, J., Kayes, A. S. M., & Watters, P. (2020). CalBehav: A machine learning-based personalized calendar behavioral model using time-series smartphone data. *The Computer Journal, 63*(7), 1109–1123.
42. Khalil, A., & Connelly, K. (2005, July). Context-aware Configuration: A study on improving cell phone awareness. In *International and interdisciplinary conference on modeling and using context* (pp. 197–209). Berlin, Heidelberg: Springer.
43. Salovaara, A., Lindqvist, A., Hasu, T., & Häkkilä, J. (2011, August). The phone rings but the user doesn't answer: Unavailability in mobile communication. In *Proceedings of the 13th International Conference on Human Computer Interaction with Mobile Devices and Services* (pp. 503–512).

44. Dekel, A., Nacht, D., & Kirkpatrick, S. (2009, September). Minimizing mobile phone disruption via smart profile management. In *Proceedings of the 11th International Conference on Human-Computer Interaction with Mobile Devices and Services* (pp. 1–5).

45. Zulkernain, S., Madiraju, P., & Ahamed, S. I. (2010, June). A context aware interruption management system for mobile devices. In *International conference on mobile wireless middleware, operating systems, and applications* (pp. 221–234). Berlin, Heidelberg: Springer.

46. Seo, S. S., Kwon, A., Kang, J. M., Strassner, J., & Hong, J. W. K. (2011, June). Pyp: Design and implementation of a context-aware configuration manager for smartphones. In *Proceedings of the 1st International Workshop on Smart Mobile Applications (SmartApps 11)*.

47. Khalil, A., & Connelly, K. (2005, September). Improving cell phone awareness by using calendar information. In *IFIP Conference on human-computer interaction* (pp. 588–600). Berlin, Heidelberg: Springer.

48. De Guzman, E. S., Sharmin, M., & Bailey, B. P. (2007, May). Should I call now? Understanding what context is considered when deciding whether to initiate remote communication via mobile devices. In *Proceedings of Graphics Interface 2007* (pp. 143–150).

49. Grandhi, S., & Jones, Q. (2010). Technology-mediated interruption management. *International Journal of Human-Computer Studies, 68*(5), 288–306.

50. Rosenthal, S., Dey, A. K., & Veloso, M. (2011, June). Using decision-theoretic experience sampling to build personalized mobile phone interruption models. In *International conference on pervasive computing* (pp. 170–187). Berlin, Heidelberg: Springer.

51. Lovett, T., O'Neill, E., Irwin, J., & Pollington, D. (2010, September). The calendar as a sensor: analysis and improvement using data fusion with social networks and location. In *Proceedings of the 12th ACM International Conference on Ubiquitous Computing* (pp. 3–12).

52. Sarker, I. H., Colman, A., Kabir, M. A., & Han, J. (2018). Individualized time-series segmentation for mining mobile phone user behavior. *The Computer Journal, 61*(3), 349–368.

53. Ozer, M., Keles, I., Toroslu, H., Karagoz, P., & Davulcu, H. (2016). Predicting the location and time of mobile phone users by using sequential pattern mining techniques. *The Computer Journal, 59*(6), 908–922.

54. Do, T. M. T., & Gatica-Perez, D. (2014). Where and what: Using smartphones to predict next locations and applications in daily life. *Pervasive and Mobile Computing, 12*, 79–91.

55. Farrahi, K., & Gatica-Perez, D. (2008, October). What did you do today? Discovering daily routines from large-scale mobile data. In *Proceedings of the 16th ACM International Conference on Multimedia* (pp. 849–852).

56. Karatzoglou, A., Baltrunas, L., Church, K., & Böhmer, M. (2012, October). Climbing the app wall: Enabling mobile app discovery through context-aware recommendations. In *Proceedings of the 21st ACM International Conference on Information and Knowledge Management* (pp. 2527–2530).

57. Phithakkitnukoon, S., Horanont, T., Di Lorenzo, G., Shibasaki, R., & Ratti, C. (2010, August). Activity-aware map: Identifying human daily activity pattern using mobile phone data. In *International workshop on human behavior understanding* (pp. 14–25). Berlin, Heidelberg: Springer.

58. Halvey, M., Keane, M. T., & Smyth, B. (2005, September). Time-based segmentation of log data for user navigation prediction in personalization. In *The 2005 IEEE/WIC/ACM international conference on web intelligence (WI'05)* (pp. 636–640). IEEE.

59. Das, G., Lin, K. I., Mannila, H., Renganathan, G., & Smyth, P. (1998, August). Rule discovery from time series. In *KDD* (Vol. 98, No. 1, pp. 16–22).

60. Lu, E. H. C., Tseng, V. S., & Philip, S. Y. (2010). Mining cluster-based temporal mobile sequential patterns in location-based service environments. *IEEE Transactions on Knowledge and Data Engineering, 23*(6), 914–927.

61. Kandasamy, K., & Kumar, C. S. (2015). Modified PSO based optimal time interval identification for predicting mobile user behaviour (MPSO-OTI2-PMB) in location based services. *Indian Journal of Science and Technology, 8*, 185.

62. Hartono, R. N., Pears, R., Kasabov, N., & Worner, S. P. (2014, July). Extracting temporal knowledge from time series: A case study in ecological data. In *2014 international joint conference on neural networks (IJCNN)* (pp. 4237–4243). IEEE.
63. Keogh, E., Chu, S., Hart, D., & Pazzani, M. (2004). Segmenting time series: A survey and novel approach. In *Data mining in time series databases* (pp. 1–21).
64. Shokoohi-Yekta, M., Chen, Y., Campana, B., Hu, B., Zakaria, J., & Keogh, E. (2015, August). Discovery of meaningful rules in time series. In *Proceedings of the 21th ACM SIGKDD International Conference on Knowledge Discovery and Data Mining* (pp. 1085–1094).
65. Phithakkitnukoon, S., Dantu, R., Claxton, R., & Eagle, N. (2011). Behavior-based adaptive call predictor. *ACM Transactions on Autonomous and Adaptive Systems, 6*(3), 1–28.
66. Jang, B. R., Noh, Y., Lee, S. J., & Park, S. B. (2015, February). A combination of temporal and general preferences for app recommendation. In *2015 international conference on big data and smart computing (BigComp)* (pp. 178–185). IEEE.
67. Henze, N., & Boll, S. (2011, August). Release your app on Sunday eve: Finding the best time to deploy apps. In *Proceedings of the 13th International Conference on Human Computer Interaction with Mobile Devices and Services* (pp. 581–586).
68. Xu, Q., Erman, J., Gerber, A., Mao, Z., Pang, J., & Venkataraman, S. (2011, November). Identifying diverse usage behaviors of smartphone apps. In *Proceedings of the 2011 ACM SIGCOMM Conference on Internet Measurement Conference* (pp. 329–344).
69. Böhmer, M., Hecht, B., Schöning, J., Krüger, A., & Bauer, G. (2011, August). Falling asleep with angry birds, facebook and kindle: A large scale study on mobile application usage. In *Proceedings of the 13th International Conference on Human Computer Interaction with Mobile Devices and Services* (pp. 47–56).
70. Xu, R., & Wunsch, D. (2005). Survey of clustering algorithms. *IEEE Transactions on Neural Networks, 16*(3), 645–678.
71. MacQueen, J. (1967, June). Some methods for classification and analysis of multivariate observations. In *Proceedings of the Fifth Berkeley Symposium on Mathematical Statistics and Probability* (Vol. 1, No. 14, pp. 281–297).
72. Rokach, L. (2009). A survey of clustering algorithms. In *Data mining and knowledge discovery handbook* (pp. 269–298). Boston, MA: Springer.
73. Sneath, P. H. (1957). The application of computers to taxonomy. *Microbiology, 17*(1), 201–226.
74. Sorenson, T. (1948). A method of establishing groups of equal amplitude in plant sociology based on similarity of species content. *K Dan Vidensk Selsk Biol Skr, 5*, 1–34.
75. Agrawal, R., & Srikant, R. (1994, September). Fast algorithms for mining association rules. In *Proc. 20th Int. Conf. Very Large Data Bases, VLDB* (Vol. 1215, pp. 487–499).
76. Quinlan, J. R. (2014). *C4. 5: Programs for machine learning*. Elsevier.
77. Agrawal, R., Imieliński, T., & Swami, A. (1993, June). Mining association rules between sets of items in large databases. In *Proceedings of the 1993 ACM SIGMOD International Conference on Management of Data* (pp. 207–216).
78. Flach, P. A., & Lachiche, N. (2001). Confirmation-guided discovery of first-order rules with Tertius. *Machine Learning, 42*(1), 61–95.
79. Houtsma, M., & Swami, A. (1995, March). Set-oriented mining for association rules in relational databases. In *Proceedings of the Eleventh International Conference on Data Engineering* (pp. 25–33). IEEE.
80. Ma, B. L. W. H. Y., Liu, B., & Hsu, Y. (1998, August). Integrating classification and association rule mining. In *Proceedings of the Fourth International Conference on Knowledge Discovery and Data Mining*.
81. Han, J., Pei, J., & Yin, Y. (2000). Mining frequent patterns without candidate generation. *ACM SIGMOD Record, 29*(2), 1–12.
82. Das, A., Ng, W. K., & Woon, Y. K. (2001, October). Rapid association rule mining. In *Proceedings of the Tenth International Conference on Information and Knowledge Management* (pp. 474–481).

83. Zhao, Q., & Bhowmick, S. S. (2003). *Association rule mining: A survey* (Vol. 135). Singapore: Nanyang Technological University.
84. Scheffer, T. (2005). Finding association rules that trade support optimally against confidence. *Intelligent Data Analysis, 9*(4), 381–395.
85. Freitas, A. A. (2000). Understanding the crucial differences between classification and discovery of association rules: A position paper. *ACM SIGKDD Explorations Newsletter, 2*(1), 65–69.
86. Fournier-Viger, P., & Tseng, V. S. (2012, December). Mining top-k non-redundant association rules. In *International symposium on methodologies for intelligent systems* (pp. 31–40). Berlin, Heidelberg: Springer.
87. Bouker, S., Saidi, R., Yahia, S. B., & Nguifo, E. M. (2012, November). Ranking and selecting association rules based on dominance relationship. In *2012 IEEE 24th international conference on tools with artificial intelligence* (Vol. 1, pp. 658–665). IEEE.
88. Han, J., Kamber, M., & Pei, J. (2011). Data mining concepts and techniques third edition. *The Morgan Kaufmann Series in Data Management Systems, 5*(4), 83–124.
89. Witten, I. H., & Frank, E. (2002). Data mining: Practical machine learning tools and techniques with Java implementations. *ACM SIGMOD Record, 31*(1), 76–77.
90. Holte, R. C. (1993). Very simple classification rules perform well on most commonly used datasets. *Machine Learning, 11*(1), 63–90.
91. Frank, E., & Witten, I. H. (1998). Generating accurate rule sets without global optimization.
92. Quinlan, J. R. (1986). Induction of decision trees. *Machine Learning, 1*(1), 81–106.
93. Lewis, R. J. (2000, May). An introduction to classification and regression tree (CART) analysis. In *Annual meeting of the society for academic emergency medicine in San Francisco, California* (Vol. 14).
94. Mehta, M., Agrawal, R., & Rissanen, J. (1996, March). SLIQ: A fast scalable classifier for data mining. In *International conference on extending database technology* (pp. 18–32). Berlin, Heidelberg: Springer.
95. Shafer, J., Agrawal, R., & Mehta, M. (1996, September). SPRINT: A scalable parallel classifier for data mining. In *Vldb* (Vol. 96, pp. 544–555).
96. Sheng, S., & Ling, C. X. (2005, October). Hybrid cost-sensitive decision tree. In *European conference on principles of data mining and knowledge discovery* (pp. 274–284). Berlin, Heidelberg: Springer.
97. Wu, X., Kumar, V., Quinlan, J. R., Ghosh, J., Yang, Q., Motoda, H. & Steinberg, D. (2008). Top 10 algorithms in data mining. *Knowledge and Information Systems, 14*(1), 1–37.
98. Wu, C. C., Chen, Y. L., Liu, Y. H., & Yang, X. Y. (2016). Decision tree induction with a constrained number of leaf nodes. *Applied Intelligence, 45*(3), 673–685.
99. Hong, J., Suh, E. H., Kim, J., & Kim, S. (2009). Context-aware system for proactive personalized service based on context history. *Expert Systems with Applications, 36*(4), 7448–7457.
100. Lee, W. P. (2007). Deploying personalized mobile services in an agent-based environment. *Expert Systems with Applications, 32*(4), 1194–1207.
101. Sarker, I. H. (2019). A machine learning based robust prediction model for real-life mobile phone data. *Internet of Things, 5*, 180–193.
102. Geng, L., & Hamilton, H. J. (2006). Interestingness measures for data mining: A survey. *ACM Computing Surveys, 38*(3), 9–es.
103. Ordonez, C. (2006, November). Comparing association rules and decision trees for disease prediction. In *Proceedings of the International Workshop on Healthcare Information and Knowledge Management* (pp. 17–24).
104. Sarker, I. H. (2019). Research issues in mining user behavioral rules for context-aware intelligent mobile applications. *Iran Journal of Computer Science, 2*(1), 41–51.
105. Sarker, I. H. (2018, March). Behavminer: Mining user behaviors from mobile phone data for personalized services. In *2018 IEEE international conference on pervasive computing and communications workshops (PerCom workshops)* (pp. 452–453). IEEE Computer Society.

106. Phithakkitnukoon, S., & Ratti, C. (2010). A recent-pattern biased dimension-reduction framework for time series data. *Journal of Advances in Information Technology, 1*(4), 168–180.
107. Sarker, I. H., Kabir, M. A., Colman, A., & Han, J. (2017, September). Understanding recency-based behavior model for individual mobile phone users. In *Proceedings of the 2017 ACM International Joint Conference on Pervasive and Ubiquitous Computing and Proceedings of the 2017 ACM International Symposium on Wearable Computers* (pp. 916–921).
108. Cheung, D. W., Han, J., Ng, V. T., & Wong, C. Y. (1996, February). Maintenance of discovered association rules in large databases: An incremental updating technique. In *Proceedings of the Twelfth International Conference on Data Engineering* (pp. 106–114). IEEE.
109. Cheung, D. W., Lee, S. D., & Kao, B. (1997). A general incremental technique for maintaining discovered association rules. In *Database systems for advanced applications' 97* (pp. 185–194).
110. Chen, J., & Shi, X. (2002). An incremental updating algorithm for mining association rules. *Computer engineering, 7.*
111. Thomas, S., Bodagala, S., Alsabti, K., & Ranka, S. (1997, August). An efficient algorithm for the incremental updation of association rules in large databases. In *KDD* (pp. 263–266).
112. Zhang, Z., Li, Y., Chen, W., & Min, F. (2014). A three-way decision approach to incremental frequent itemsets mining. *Journal of Information & Computational Science, 11*(10), 3399–3410.
113. Li, Y., Zhang, Z. H., Chen, W. B., & Min, F. (2017). TDUP: An approach to incremental mining of frequent itemsets with three-way-decision pattern updating. *International Journal of Machine Learning and Cybernetics, 8*(2), 441–453.
114. Yao, Y. (2012, August). An outline of a theory of three-way decisions. In *International conference on rough sets and current trends in computing* (pp. 1–17). Berlin, Heidelberg: Springer.
115. Amornchewin, R., & Kreesuradej, W. (2009). Mining dynamic databases using probability-based incremental association rule discovery algorithm. *Journal of UCS, 15*(12), 2409–2428.
116. Thusaranon, P., & Kreesuradej, W. (2015). A probability-based incremental association rule discovery algorithm for record insertion and deletion. *Artificial Life and Robotics, 20*(2), 115–123.
117. Lee, S., Seo, J., & Lee, G. (2010, April). An adaptive speed-call list algorithm and its evaluation with ESM. In *Proceedings of the SIGCHI Conference on Human Factors in Computing Systems* (pp. 2019–2022).
118. Barzaiq, O. O., & Loke, S. W. (2011). Adapting the mobile phone for task efficiency: The case of predicting outgoing calls using frequency and regularity of historical calls. *Personal and Ubiquitous Computing, 15*(8), 857–870.
119. Phithakkitnukoon, S., & Dantu, R. (2008). Adequacy of data for characterizing caller behavior. In *Proceedings of KDD Inter. Workshop on Social Network Mining and Analysis (SNAKDD 2008).*
120. Bergman, O., Komninos, A., Liarokapis, D., & Clarke, J. (2012). You never call: Demoting unused contacts on mobile phones using DMTR. *Personal and Ubiquitous Computing, 16*(6), 757–766.
121. Stefanis, V., Plessas, A., Komninos, A., & Garofalakis, J. (2014). Frequency and recency context for the management and retrieval of personal information on mobile devices. *Pervasive and Mobile Computing, 15*, 100–112.
122. Sarker, I. H., Colman, A., Kabir, M. A., & Han, J. (2016). Behavior-oriented time segmentation for mining individualized rules of mobile phone users. In *2016 IEEE international conference on data science and advanced analytics (DSAA)* (pp. 488–497). IEEE.
123. El Khaddar, M. A., & Boulmalf, M. (2017). Smartphone: The ultimate IoT and IoE device. *Smartphones from an Applied Research Perspective, 137.*
124. Sarker, I. H., Hoque, M. M., Uddin, M. K., & Alsanoosy, T. (2021). Mobile data science and intelligent apps: Concepts, AI-based modeling and research directions. *Mobile Networks and Applications, 26*(1), 285–303.

125. Pejovic, V., & Musolesi, M. (2014, September). InterruptMe: Designing intelligent prompting mechanisms for pervasive applications. In *Proceedings of the 2014 ACM International Joint Conference on Pervasive and Ubiquitous Computing* (pp. 897–908).

126. Peng, M., Zeng, G., Sun, Z., Huang, J., Wang, H., & Tian, G. (2018). Personalized app recommendation based on app permissions. *World Wide Web, 21*(1), 89–104.

127. Zheng, P., & Ni, L. M. (2006). Spotlight: The rise of the smart phone. *IEEE Distributed Systems Online, 7*(3), 3.

128. Google trends (2019). https://trends.google.com/trends/

129. Finin, T., Joshi, A., Kagal, L., Ratsimore, O., Korolev, V., & Chen, H. (2001, September). Information agents for mobile and embedded devices. In *International workshop on cooperative information agents* (pp. 264–286). Berlin, Heidelberg: Springer.

130. de Almeida, D. R., de Souza Baptista, C., da Silva, E. R., Campelo, C. E., de Figueirêdo, H. F., & Lacerda, Y. A. (2006, April). A context-aware system based on service-oriented architecture. In *20th international conference on advanced information networking and applications-volume 1 (AINA'06)* (Vol. 1, pp. 6-pp). IEEE.

131. Sarker, I. H. (2021). Machine learning: Algorithms, real-world applications and research directions. *SN Computer Science, 2*(3), 1–21.

132. Shi, Y. (2006, August). Context awareness, the spirit of pervasive computing. In *2006 first international symposium on pervasive computing and applications* (pp. 6–6). IEEE.

133. Anagnostopoulos, C., Tsounis, A., & Hadjiefthymiades, S. (2005, July). Context management in pervasive computing environments. In *ICPS'05. Proceedings. International Conference on Pervasive Services, 2005* (pp. 421–424). IEEE.

134. Zhu, H., Chen, E., Xiong, H., Yu, K., Cao, H., & Tian, J. (2014). Mining mobile user preferences for personalized context-aware recommendation. *ACM Transactions on Intelligent Systems and Technology, 5*(4), 1–27.

135. Sarker, I. H., Colman, A., Kabir, M. A., & Han, J. (2016, September). Phone call log as a context source to modeling individual user behavior. In *Proceedings of the 2016 ACM International Joint Conference on Pervasive and Ubiquitous Computing: Adjunct* (pp. 630–634).

136. Eagle, N., & Pentland, A. S. (2006). Reality mining: Sensing complex social systems. *Personal and Ubiquitous Computing, 10*(4), 255–268.

137. Srinivasan, V., Moghaddam, S., Mukherji, A., Rachuri, K. K., Xu, C., & Tapia, E. M. (2014, September). Mobileminer: Mining your frequent patterns on your phone. In *Proceedings of the 2014 ACM International Joint Conference on Pervasive and Ubiquitous Computing* (pp. 389–400).

138. Mehrotra, A., Hendley, R., & Musolesi, M. (2016, September). PrefMiner: Mining user's preferences for intelligent mobile notification management. In *Proceedings of the 2016 ACM International Joint Conference on Pervasive and Ubiquitous Computing* (pp. 1223–1234).

139. Paireekreng, W., Rapeepisarn, K., & Wong, K. W. (2009). Time-based personalised mobile game downloading. In *Transactions on edutainment II* (pp. 59–69). Berlin, Heidelberg: Springer.

140. Cao, H., Bao, T., Yang, Q., Chen, E., & Tian, J. (2010, October). An effective approach for mining mobile user habits. In *Proceedings of the 19th ACM International Conference on Information and Knowledge Management* (pp. 1677–1680).

141. Rawassizadeh, R., Tomitsch, M., Wac, K., & Tjoa, A. M. (2013). UbiqLog: A generic mobile phone-based life-log framework. *Personal and Ubiquitous Computing, 17*(4), 621–637.

142. Cao, L. (2017). Data science: A comprehensive overview. *ACM Computing Surveys, 50*(3), 1–42.

143. Haghighi, P. D., Krishnaswamy, S., Zaslavsky, A., Gaber, M. M., Sinha, A., & Gillick, B. (2013). Open mobile miner: A toolkit for building situation-aware data mining applications. *Journal of Organizational Computing and Electronic Commerce, 23*(3), 224–248.

Part II
Context-Aware Rule Learning and Management

This part of the book explores the approaches to extract the context-aware rules from mobile data in multi-dimensional contexts by presenting the datasets and contextual features (Chap. 4), the approach to discretization of time-series data (Chap. 5), the approach to extract rules (Chap. 6), and the approach to managing the recent patterns (Chap. 7).

Chapter 4
Contextual Mobile Datasets, Pre-processing and Feature Selection

4.1 Smart Mobile Phone Data and Associated Contexts

We live in the data era [1], in which everything around us is connected to a data source, and everything in our lives is recorded digitally. Mobile phones, also known as cellular phones, have become increasingly common and popular. Their ability to capture user behaviors allows them to gain insight into individual users' cell phone use habits. Recent advancements in smart cell phones and their sensing capabilities have allowed the collection of rich contextual information about the user and mobile phone usage records through system logs, such as phone call logs [2, 3], SMS Log [4], mobile application (Apps) usages logs [5, 6], mobile phone notification logs [7], weblogs [8], Game Log [9], context logs [5], and smartphone lifelog [10] etc. These historical mobile phone data is simply a combination of past contextual information (various contexts) and individual mobile phone users' behaviors for those contexts [11]. As a result, such contextual data captured by cell phones can be used as the basis for analyzing user behavioral activities. In the following, we discuss several phone log datasets and their associated contexts.

4.1.1 Phone Call Log

Smartphones can store different types of contextual data related to a user's phone call activities, such as making outgoing calls and handling incoming calls, in its phone log. The logging of a user's phone call information, e.g., call date, exact call time, call type, call duration in call logs, in particular, provides raw data on when the user makes outgoing calls, accepts, declines, or misses incoming calls [3, 12, 13]. In addition to call specific metadata, other types of contextual information, such as user location and the social relationship between the caller and the callee identified by the individual's unique phone contact number, are also registered by smart cell

© The Author(s), under exclusive license to Springer Nature Switzerland AG 2021
I. H. Sarker et al., *Context-Aware Machine Learning and Mobile Data Analytics*,
https://doi.org/10.1007/978-3-030-88530-4_4

phones [4]. Thus, call log data obtained by a smart cell phone can be used as a contextual data source in smart context-aware mobile communication systems to model individual mobile phone user behavior [14].

4.1.2 Mobile SMS Log

The most popular text communication service in cell phone communication systems is the short message service (SMS). Individual cell phone users can exchange short text messages using uniform communications rules or protocols. According to the International Telecommunication Union's [15], short messages have developed into a vast commercial industry worth over 81 billion dollars. The rapid increase in the number of cell phone users around the world has resulted in a significant increase in spam messages. The main issue with SMS spam is that it is irritating to the recipients. Furthermore, SMS spam can be costly due to some users having to pay for each text message they receive. Furthermore, permanently blocking SMS from a specific subscriber is not a good option because of a possibility to miss a significant message. The SMS log contains all messages, including real and spam text messages [16], or good content and bad content [17], as well as their contextual information, such as the user identifier, date, time, and other SMS related metadata, which can be used in the task of automatic filtering of SMS spam for different individuals in different contexts [4, 16], or predicting good time or bad time to deliver such messages [17].

4.1.3 Smartphone App Usage Log

Due to the rapid growth and adoption of mobile channels, smartphones and tablets have become one of the most significant media for social entertainment and knowledge acquisition [5]. The vast number of mobile applications, or apps, e.g., Multimedia, Facebook, Gmail, Youtube, Skype, Game available via the Internet, which can be installed and operated by individual users on their smartphones according to their individual needs, can be linked to the radical rise of smart mobile phones. As a result, it's important for not only mobile designers and software developers, but also academics, to understand how different people use various applications for real-world purposes on their smartphones. For different types of mobile apps, app use logs contain various contextual information such as date, time of day, battery level, profile type such as general, silent, meeting, outdoor, offline, charging state such as charging, full, or not connected, location such as home, workplace, on the way, and other app-related metadata [5, 6, 18, 19]. These logs can be used to mine the contextual behavioral rules of individual cell phone users, such as which app a user prefers in a given situation. Mining such preferences is a crucial step toward better understanding mobile phone users' app use patterns and providing customized context-aware app recommendations according to their current needs.

4.1.4 Mobile Phone Notification Log

In the real world, several smart mobile apps use notifications to notify users of different events, news, or simply to give them reminders or warnings. Many of them, for example, mobile phone alerts are neither useful nor important to the users' interests. As a consequence, such ineffective alerts are regarded as obtrusive and potentially irritating by users [7]. For instance, notifications of inviting games on social networks, social or promotional emails, or a variety of predictive suggestions provided by various smartphone applications, such as Twitter, Facebook, LinkedIn, WhatsApp, Viver, Skype, and Youtube [7]. The notification log contains contextual data such as notification type, user physical activity such as still, walking, running, biking, or driving, user location such as home or work place, date, time-of-day, user reaction to such notifications, dismiss or accept, and other notification-related metadata [7]. Such notification logs can be used to extract contextual behavioral rules that can be used to create intelligent mobile notification management systems according to their preferences.

4.1.5 Web or Navigation Log

Individual users may have different needs and priorities at different times of the day and days of the week because user navigation habits on the Internet are context-dependent [8]. In recent years, there has been a growing interest in time-based consumption trends in the workplace. User mobile web navigation, web browsing, e-mail, entertainment, talk, news, TV, travel, sport, banking, miscellaneous, and related contextual information such as date, time-of-day, weekdays, weekends are all recorded in the weblog [8, 20, 21]. Mining contextual usage patterns from log data can be used to make accurate context-aware navigation predictions and adjust the portal layout to the needs of users. Such contextual user navigation predictions can help to improve download times by pre-fetching pages, which would be beneficial for mobile phone users for whom download time is a particular issue.

4.1.6 Game Log

With the rapid development and widespread availability of mobile Internet, smartphones are now being used for a variety of traditional PC-based applications, such as game downloads. Downloading mobile games via smartphone appears to be one of the most common activities among mobile phone users these days. People enjoy a wide variety of games, including action, adventure, casual, puzzle, strategy, and sports. Users of mobile phones can play such games for a variety of reasons, including to relax after a challenge, to quickly escape from real-world activities, or to pass the time when they are lonely. Users can, however, have different preferences about what types of games they want to play at different times of the day. Individual cell phone users' game logs provide information about playing different types of games, as well as relevant contextual information such as date, time of day, weekdays, weekends, and so on [9]. The context-dependent game-playing rules derived from such log data can be used to create a customized mobile game recommendation system for individual phone users based on their preferences.

4.1.7   Smartphone Life Log

Smartphones are more than just a phone for making calls; they also have video players, game consoles, personal calendars, storage, and other advanced features. They are also known as handheld computers, and they have the same computing capabilities as personal computers. Individual users, on the other hand, can take their smart mobile phones with them anywhere and at any time, unlike personal computers. As a result of the widespread availability of smart cell phones and their computational capacities for a variety of real-world applications, these devices can be used as a life-logging system, such as personal e-memories [10]. In a more technological sense, life-logs sense and store contextual information from an individual's surrounding environment using a range of sensors found in their smartphones, which are the life-long core components such as user phone calls, SMS headers, App use, e.g., Skype, Whatsapp, Youtube etc., physical activities form Google play API, and related contextual information such as WiFi and Bluetooth devices in user's proximity, geographical location, temporal information [10]. Individual mobile phone users' context-dependent behavioral rules can be derived from such life-log data and used to enhance user experience in their everyday lives.

4.1.8   Dataset Summary

In the above, we have discussed various types of smartphone log data related to the user's activities and their associated contextual information. Hence, we summarize the major characteristics of such log data that is collected by the smart mobile phones in a context-aware pervasive computing environment.

- Data are collected automatically by the devices according to their data storage capabilities, thus minimized the effort of manual data collection.

- These are the real-life data of mobile phone users with their unique behavioral patterns, thus contains the actual activities or response with the devices, e.g., reject phone call, of the individual users in different contexts such as temporal, spatial, social or others relevant.
- Different types of customized apps can be used to record the necessary data for building data-driven personalized applications for the mobile phone users.

This represents the first layer of our context-aware rule learning framework as collecting relevant data is the first step to build a data-driven system. Such contextual data can be collected from various sources like smartphone logs, sensors or external sources relevant to the application. Smartphone data collected from these sources usually contains raw contexts that characterize individuals' daily life behavioral activities with their phones, and need to process effectively to use as the basis for learning context-aware rules.

4.2 Examples of Contextual Mobile Phone Data

In this section, several data samples in different contexts have been provided to give a clear understanding on contextual datasets. This includes time-series mobile phone data as well as data with multi-dimensional contexts for different applications.

4.2.1 Time-Series Mobile Phone Data

Let, T_{series} be a data feature and q it's corresponding domain. T_{series} is a sequence of data points ordered in time such that $T_{series} = (t_1, t_2, ..., t_m)$, where $t_1, t_2, ..., t_m$ are individual observations, each of which contains real-value data and m is the number of observations in a time-series. A time-series mobile phone dataset DS_T is a collection of records, where

(i) each record r is a set of pairs $(t_i, value_i)$, where $t_i \in T_{series}$ that represents the timestamps, and $value \in q$. For example, '2016-09-10 19:38:20' is a value of t_i, which represents the timestamps information in the format YYYY-MM-DD hh:mm:ss.
(ii) each $t_i \in T_{series}$, also called attribute (temporal context), may occur at most once in any record, and
(iii) each record has a particular user activity with mobile phones (e.g., reject phone call).

A sample user's time-series cell phone data is shown in Table 4.1. It presents some data from a phone call log, which tracks user phone call activities along with corresponding temporal data. In the real world, an individual cell phone user's typical phone call habits are (i) accepting an incoming phone call, (ii) rejecting

Table 4.1 Examples of time-series mobile phone data and corresponding phone call activities of a sample mobile phone user

Timestamps (YYYY-MM-DD hh:mm:ss)	Phone call type and duration	User phone call activity (behavior)
2016-09-19 10:03:15	Incoming call, 0 s	Reject
2016-09-19 10:35:25	Incoming call, 0 s	Reject
2016-09-19 11:53:55	Incoming call, 0 s	Reject
2016-09-19 21:30:22	Outgoing call, 125 s	Outgoing
2016-09-19 22:40:25	Incoming call, 125 s	Accept
2016-09-20 07:53:35	Missed call, 0 s	Missed
2016-09-20 08:19:14	Outgoing call, 60 s	Outgoing
2016-09-20 08:25:04	Incoming call, 135 s	Accept
2016-09-20 10:19:39	Incoming call, 0 s	Reject
2016-09-20 12:20:24	Missed call, 0 s	Missed

or declining an incoming phone call, (iii) the phone rings but the user ignores the call intentionally or unintentionally, i.e, missed, and (iv) making a phone call to a specific person, i.e., Outgoing [3]. Both temporal information and user-related phone call activity are stored for each record, with accept and reject calls being stored in the system log as incoming calls with corresponding call duration [3].

Each record in the dataset (see Table 4.1) contains exact temporal information (e.g., 2016-09-19 10:03:15) that can not be used in their behavioral rules as there is a very small chance that these exact values may match values of unseen, testing cases to predict her future behavior. For instance, the first record shows that the user rejects the incoming phone call on 2016-09-19 at 10:03:15. It is unlikely to reject another incoming call at the same time 10:03:15 (hh:mm: ss) on another date (YYYY-MM-DD). In general, discretization or time-series segmentation approach could be an effective approach to pre-process such temporal contextual data, in which an input is divided into a sequence of discrete segments or nominal attributes, that is, quantitative data into qualitative data, with a finite number of intervals.

4.2.2 Mobile Phone Data with Multi-Dimensional Contexts

Let $Con = \{con_1, con_2, ..., con_m\}$ be a set of nominal contexts and $Q = \{q_1, q_2, ..., q_m\}$ the set of corresponding domains. A mobile phone dataset DS is a collection of records, where

(i) each record r is a set of pairs $(con_i, value_i)$, where $con_i \in Con$, and $value_i \in Q$. For example, if con_i represents as the context 'location', then an example of $value_i$ is 'office'.

(ii) each $con_i \in Con$, also called attribute (context), may occur at most once in any record, and

Table 4.2 Sample mobile phone dataset with multi-dimensional contexts and corresponding phone call activities of a sample user

Temporal	Location	Social activity (situation)	Social relationship	User phone call activity (behavior)
Mon[10:00-12:00]	Office	Meeting	Friend	Reject
Mon[10:00-12:00]	Office	Meeting	Boss	Accept
Mon[10:00-12:00]	Office	Meeting	Unknown	Reject
Mon[10:00-12:00]	Office	Meeting	Colleague	Reject
Mon[18:30-19:30]	Home	Dinner	Friend	Accept
Mon[18:30-19:30]	Home	Dinner	Colleague	Accept
Tue[18:30-19:30]	Home	Dinner	Unknown	Missed
Tue[12:30-13:30]	Office	Lunch	Friend	Accept
Wed[15:00-17:00]	Office	Lecture	Friend	Missed
Wed[15:00-17:00]	Office	Lecture	Friend	Reject
Wed[15:00-17:00]	Office	Lecture	Friend	Reject
Fri[10:15-11:30]	Office	Seminar	Friend	Reject
Fri[10:15-11:30]	Office	Seminar	Mother	Accept
Fri[10:15-11:30]	Office	Seminar	Friend	Reject
Fri[14:30-15:30]	Office	Seminar	Friend	Reject
Fri[14:30-15:30]	Office	Seminar	Mother	Accept
Fri[14:30-15:30]	Office	Seminar	Friend	Reject
Sat[18:30-19:30]	Market	Shopping	Friend	Missed
Sat[18:30-19:30]	Market	Shopping	Unknown	Missed
Sun[18:30-19:30]	Home	Dinner	Colleague	Accept

(iii) each record has a particular user activity with mobile phones (e.g., reject phone call).

Table 4.2 shows an example of mobile phone data containing multi-dimensional contexts of a sample mobile phone user. It reports some pieces of information including user phone call activities with multi-dimensional contextual information.

Each record in the dataset (see Table 4.2) contains several contextual information related to her particular phone call activity. For instance, the first record shows that the user rejects (phone call activity) her friend's (social relationship) phone call between 10:00 and 12:00 on Monday (temporal context) as she is in a meeting (social activity) at her office (spatial context). User behavioral rules are discovered from such contextual data, that can be used to predict her future behavior in different contexts. In the following, we discuss multi-dimensional contexts for another application domain.

4.2.3   Contextual Apps Usage Data

Contextual apps usage data, discussed in this section, includes not only the user-centric context, such as the users' spatio-temporal context, their mood or desires, and so on, but also the device-centric context, which considers the users' effects on their use. It contains different types of apps usage including Facebook, Gmail, LinkedIn, Instagram , Youtube, Whatsapp, Read News etc. of a sample user. In the following, we discuss the apps usage data considering several contexts.

Temporal Context This is one of the most important factors influencing a user's smartphone use [22]. For example, an individual's smartphone app usage in the morning can differ from her usage at night. Furthermore, in the real world, one's behavioral activities can vary over different periods or hours.

Work Status In general, an individual's work status in the real world is determined by the day of the week, whether it is a workday or a holiday. Many people's work status has a major effect on their app use. For example, one's app use habits on a Saturday, say a holiday, can differ from those on other workdays.

Spatial Context It represents users' spatial information, such as their location, which can be used to model and predict individual smartphone apps in another context. The explanation for this is that an individual's phone use can be varied according to her current location [14]. As a result of this perception of user mobility and the associated context-aware model, location-based services can be provided for the benefit of individual users.

User Mood Typically, user mood refers to a person's emotional state, which is primarily concerned with sentiment and emotional analysis. Since a person's emotional condition is not constant in the real world and may shift over time, it could be another important factor that influences individuals and can be used to model customized app use conduct [23]. For example, when one person is in a good mood, she prefers to listen to music, while when she is sad, she prefers to use online messaging.

Device Status Individuals' device-specific contextual information such as phone profile, phone battery level, or charging status, in addition to the above contexts related to users' day-to-day circumstances and preferences, which affect individuals' preferences to use smartphone apps. For example, if a person's phone battery gives off a low-power signal, she is unlikely to connect to the Internet to use an entertainment app.

Internet Connectivity This often reflects the device's meaning, which connects it to the rest of the world. As a result, Internet access and speed can affect how people use their smartphones. For example, if Wifi (wireless fidelity, which primarily refers to some types of wireless local area networks) is available, one person enjoys playing video songs; otherwise, he does not.

Table 4.3 Sample examples of contextual apps usage data

Contexts	Type	Example values
Temporal context	Continuous	Time-of-the-day [24-hours-a-day] Days-of-the-week [7-days-a-week]
Spatial context	Categorical	Phone user location [at home, at office, at the canteen, in the playground, on the way, etc.]
Work status context	Categorical (binary)	Workday and Holiday
User mood context	Categorical	Emotional state of phone user [normal, happy, or sad]
Device status context	Categorical	Battery level [low, medium, or full]
Phone profile context	Categorical	Phone notification [general, silent, or vibration]
Internet connectivity context	Categorical (binary)	WiFi connectivity [on, off]
Smartphone apps	Categorical	Social networking, Gmail, Communication, video, entertainment, read news, games etc.

Table 4.3 provides a summary with the examples of contextual apps usage data. Extracting contextual behavioral rules of mobile phone users can be used to build an effective context-aware model, such as an intelligent apps recommendation system, that effectively predicts personalized smartphone app usage and recommends accordingly in their daily usage, based on the contextual information discussed above.

4.3 Data Preprocessing

This section covers the basics of data pre-processing, such as eliminating noisy data, replacing missing values, and selecting important features while removing unnecessary and redundant ones. The followings are included in the pre-processing step:

4.3.1 Data Cleaning

The quality of the data is important to build a data-driven model. Any data that is incomplete, noisy, or inconsistent may have an impact on the final outcome [24]. Data cleaning is the method of deleting or altering data that is inaccurate, incomplete, obsolete, duplicated, or incorrectly formatted in order to prepare it for further processing. When it comes to data analysis, this data is normally not required or beneficial because it can slow down the process or produce incorrect results.

Missing values can be handled by ignoring the tuple or filling the value with a specific value. Noisy data can be handled in many ways such as binning techniques, clustering, combined human and machine analysis, regression, etc. Inconsistency can be resolved manually. Overall, we can conclude that data cleaning process prepares the data for further processing, which extracts the most useful information from the collected data.

4.3.2 Data Integration

In certain cases, we need to integrate data from different sources in a data warehouse for further study [25]. Knowledge silos are typically created due to the lack of collaboration, making it difficult to get a full picture of how an organization is doing. It causes inefficiencies, slowing decision-making and increasing redundancies. Data integration is a data pre-processing technique that involves integrating data from several heterogeneous data sources into a single data store. Data can be made more important with the aid of a good integration strategy. In data integration, schema integration and redundancy are major concerns. Thus, we must deal with a number of issues when integrating the data, including entity identification problems, redundancy, tuple duplication, and data conflict detection and resolution.

4.3.3 Data Transformation

The process of transforming data from one format or structure to another is known as data transformation in computing. Most data integration and data management functions, including data wrangling, data warehousing, data integration, and device integration, require it. Data transformation can be simple or complicated, depending on the amount of data that needs to be changed between the source and target data. The format, structure, complexity, and volume of the data being transformed can all influence the tools and technologies used for data transformation. It is usually accomplished by a combination of manual and automated processes. A few transformation approaches include normalization, smoothing, discretization, aggregation, and generalization can be used. However, during transition, a lack of experience and carelessness can cause issues.

4.3.4 Data Reduction

Data reduction is a method of reducing the size of original data so that it can be represented in a much smaller space. When reducing data, data reduction strategies maintain data integrity. The time spent on data reduction should not be overlooked in favor of the time saved by data mining on the smaller data collection. Data reduction can serve two purposes: it can minimize the number of data records by removing invalid data, or it can generate summary data and statistics at various aggregation levels for various applications. Data aggregation, attribute subset selection, dimension reduction, data compression, numerosity reduction, discretization, hierarchy generation, etc. are all methods that can be used. Data reduction is performed after data cleaning, incorporation, and transformation to obtain task-relevant data.

Pre-processing is necessary when the data set contains irrelevant data that is incomplete (missing), noisy (outliers), and inconsistent. Thus, in the field of data analytics, pre-processing is an important and prerequisite phase that is used to turn raw data into a useful and efficient format. In the following, we discuss several types of feature selection as well as dimensionality reduction techniques.

4.4 Dimensionality Reduction

High-dimensional data processing is a difficult task for both researchers and application developers in the area of machine learning and data science. Thus, unsupervised learning technique dimensionality reduction is significant because it leads to better human interpretations, lower computational costs, and avoids over-fitting and duplication by simplifying models. Feature selection is a process to select features which are more informative but some features may be redundant, and others may be irrelevant and noisy [26]. For dimensionality reduction, both feature selection and feature extraction can be used. The main difference between these two is that feature selection selects a subset of the original features, while feature extraction produces entirely new ones [26]. In the following, we briefly discuss these techniques.

4.4.1 Feature Selection

The process of choosing a subset of specific features (variables, predictors) to use in building machine learning and data science models is known as feature selection, also known as variable or attribute selection in data. It reduces the complexity of a model by removing irrelevant or less important features, allowing machine learning algorithms to be trained faster. A correct and optimal subset of the selected features in a problem domain will reduce overfitting by simplifying and generalizing the model while also increasing the model's accuracy [27]. As a result, feature selection [28, 29] is regarded as one of the most important concepts in machine learning, as it has a significant impact on the target machine learning model's effectiveness and performance. Some common techniques for feature selection include the Chi-squared test, analysis of variance (ANOVA), Pearson's correlation coefficient, and recursive feature elimination.

4.4.2   Feature Extraction

Feature extraction techniques in a machine learning-based model or system typically provide a better understanding of the data, a way to increase prediction accuracy, and a way to minimize computational cost or training time. The aim of feature extraction [28, 29] is to reduce the number of features in a dataset by creating new ones from old ones and discarding the old ones. This new reduced set of features can then be used to summarize the original set of features. Principal components analysis (PCA) is a dimensionality-reduction technique that extracts a lower-dimensional space from existing features in a dataset to create new brand components [29].

4.4.3   Dimensionality Reduction Algorithms

In the machine learning and data science literature, several algorithms have been suggested to minimize data dimensions [30, 31]. The common methods that are commonly used in different application areas are summarized in the following sections.

- *Chi-square:* The chi-square χ^2 [32] statistic estimates the discrepancy between observed and predicted frequencies of a sequence of events or variables. The degree of freedom, the sample size, and the magnitude of the disparity between the actual and observed values are all influenced by χ^2. For evaluating relationships between categorical variables, the chi-square χ^2 is widely used. If E_i is the predicted value and O_i is the observed value, then

$$\chi^2 = \sum_{i=1}^{n} \frac{(O_i - E_i)^2}{E_i} \tag{4.1}$$

- *Pearson Correlation:* Another tool for understanding a feature's relationship to the response variable is Pearson's correlation, which can be used for feature selection [29]. This approach can also be used to determine the relationship between features in a dataset. The resulting value is $[-1, 1]$, where -1 means perfect negative correlation, $+1$ means perfect positive correlation, and 0 means that the two variables do not have a linear correlation. If X and Y are two random variables, the correlation coefficient between X and Y is defined as [30]

$$r(X, Y) = \frac{\sum_{i=1}^{n}(X_i - \bar{X})(Y_i - \bar{Y})}{\sqrt{\sum_{i=1}^{n}(X_i - \bar{X})^2}\sqrt{\sum_{i=1}^{n}(Y_i - \bar{Y})^2}} \tag{4.2}$$

- *ANOVA:* ANOVA is a statistical method for comparing the mean values of two or more groups that are substantially different from one another. ANOVA assumes that the variables and the objective have a linear relationship and that the variables have a normal distribution. F-tests are used in the ANOVA system to statistically verify the equality of means. The results of this test's ANOVA F-value' [32] can be used for feature selection when some features independent of the target variable can be omitted.

- *Recursive Feature Elimination (RFE):* Recursive Feature Elimination (RFE) is a feature selection method that uses brute force. Before the model meets the required number of features, RFE [32] suits it and eliminates the weakest function. Features are ranked by the coefficients or feature significance of the model. By recursively extracting a small number of features per iteration, RFE aims to eliminate model dependencies and collinearity.
- *Principal Component Analysis (PCA):* In the field of machine learning and data science, principal component analysis (PCA) is a widely used unsupervised learning method. PCA is a mathematical technique that converts a collection of correlated variables into a set of uncorrelated variables called principal components from a set of correlated variables [33, 34]. PCA works by identifying the fully transformed covariance matrix with the highest eigenvalues and then projecting the data into a new subspace of equal or fewer dimensions [32].

4.5 Conclusion

Several contextual datasets captured by the mobile devices that can be utilized to extract contextual rules for building rule-based context-aware model are presented in this chapter. Such contextual data can be collected from various sources like smartphone logs, sensors or external sources relevant to the application. Smartphone data collected from these sources usually contains raw contexts that characterize individuals' daily life behavioral activities with their phones, and need to process effectively to use as the basis for learning context-aware rules. As the real-world data may contain noisy and inconsistency instances, the pre-processing steps have also been analyzed to clean and remove noises from raw data, which is an important and prerequisite phase that is used to turn raw data into a useful and efficient format. Finally, the basic feature selection and extraction methods for efficient processing has also been provided in this chapter

References

1. Cao, L. (2017). Data science: A comprehensive overview. *ACM Computing Surveys, 50*(3), 1–42.
2. Phithakkitnukoon, S., Dantu, R., Claxton, R., & Eagle, N. (2011). Behavior-based adaptive call predictor. *ACM Transactions on Autonomous and Adaptive Systems, 6*(3), 1–28.
3. Sarker, I. H., Colman, A., Kabir, M. A., & Han, J. (2016, September). Phone call log as a context source to modeling individual user behavior. In *Proceedings of the 2016 ACM International Joint Conference on Pervasive and Ubiquitous Computing: Adjunct* (pp. 630–634).

4. Eagle, N., & Pentland, A. S. (2006). Reality mining: Sensing complex social systems. *Personal and Ubiquitous Computing, 10*(4), 255–268.
5. Zhu, H., Chen, E., Xiong, H., Yu, K., Cao, H., & Tian, J. (2014). Mining mobile user preferences for personalized context-aware recommendation. *ACM Transactions on Intelligent Systems and Technology, 5*(4), 1–27.
6. Srinivasan, V., Moghaddam, S., Mukherji, A., Rachuri, K. K., Xu, C., & Tapia, E. M. (2014, September). Mobileminer: Mining your frequent patterns on your phone. In *Proceedings of the 2014 ACM International Joint Conference on Pervasive and Ubiquitous Computing* (pp. 389–400).
7. Mehrotra, A., Hendley, R., & Musolesi, M. (2016, September). PrefMiner: Mining user's preferences for intelligent mobile notification management. In *Proceedings of the 2016 ACM International Joint Conference on Pervasive and Ubiquitous Computing* (pp. 1223–1234).
8. Halvey, M., Keane, M. T., & Smyth, B. (2005, September). Time-based segmentation of log data for user navigation prediction in personalization. In *The 2005 IEEE/WIC/ACM international conference on web intelligence (WI'05)* (pp. 636–640). IEEE.
9. Paireekreng, W., Rapeepisarn, K., & Wong, K. W. (2009). Time-based personalised mobile game downloading. In *Transactions on edutainment II* (pp. 59–69). Berlin, Heidelberg: Springer.
10. Rawassizadeh, R., Tomitsch, M., Wac, K., & Tjoa, A. M. (2013). UbiqLog: A generic mobile phone-based life-log framework. *Personal and Ubiquitous Computing, 17*(4), 621–637.
11. Hong, J., Suh, E. H., Kim, J., & Kim, S. (2009). Context-aware system for proactive personalized service based on context history. *Expert Systems with Applications, 36*(4), 7448–7457.
12. Bell, S., McDiarmid, A., & Irvine, J. (2011, May). Nodobo: Mobile phone as a software sensor for social network research. In *2011 IEEE 73rd vehicular technology conference (VTC Spring)* (pp. 1–5). IEEE.
13. Pielot, M. (2014, September). Large-scale evaluation of call-availability prediction. In *Proceedings of the 2014 ACM International Joint Conference on Pervasive and Ubiquitous Computing* (pp. 933–937).
14. Sarker, I. H., & Kayes, A. S. M. (2020). ABC-RuleMiner: User behavioral rule-based machine learning method for context-aware intelligent services. *Journal of Network and Computer Applications, 168*, 102762.
15. International Telecommunication Union, ITU Internet Report, 2006.
16. Almeida, T. A., Hidalgo, J. M. G., & Yamakami, A. (2011, September). Contributions to the study of SMS spam filtering: New collection and results. In *Proceedings of the 11th ACM Symposium on Document Engineering* (pp. 259–262).
17. Fischer, J. E., Yee, N., Bellotti, V., Good, N., Benford, S., & Greenhalgh, C. (2010, September). Effects of content and time of delivery on receptivity to mobile interruptions. In *Proceedings of the 12th International Conference on Human Computer Interaction with Mobile Devices and Services* (pp. 103–112).
18. Kim, J., & Mielikäinen, T. (2014, September). Conditional log-linear models for mobile application usage prediction. In *Joint European conference on machine learning and knowledge discovery in databases* (pp. 672–687). Berlin, Heidelberg: Springer.
19. Zhu, H., Chen, E., Xiong, H., Cao, H., & Tian, J. (2013). Mobile app classification with enriched contextual information. *IEEE Transactions on mobile computing, 13*(7), 1550–1563.
20. Halvey, M., Keane, M. T., & Smyth, B. (2006, April). Time based patterns in mobile-internet surfing. In *Proceedings of the SIGCHI Conference on Human Factors in Computing Systems* (pp. 31–34).
21. Bordino, I., Donato, D., & Poblete, B. (2012, October). Extracting interesting association rules from toolbar data. In *Proceedings of the 21st ACM International Conference on Information and Knowledge Management* (pp. 2543–2546).
22. Sarker, I. H., Colman, A., Kabir, M. A., & Han, J. (2018). Individualized time-series segmentation for mining mobile phone user behavior. *The Computer Journal, 61*(3), 349–368.

23. Sarker, I. H., & Salah, K. (2019). Appspred: Predicting context-aware smartphone apps using random forest learning. *Internet of Things, 8,* 100106.
24. Sarker, I. H. (2019). A machine learning based robust prediction model for real-life mobile phone data. *Internet of Things, 5,* 180–193.
25. Sarker, I. H., Colman, A., Han, J., Kayes, A. S. M., & Watters, P. (2020). CalBehav: A machine learning-based personalized calendar behavioral model using time-series smartphone data. *The Computer Journal, 63*(7), 1109–1123.
26. Sarker, I. H. (2021). Machine learning: Algorithms, real-world applications and research directions. *SN Computer Science, 2*(3), 1–21.
27. Sarker, I. H., Abushark, Y. B., Alsolami, F., & Khan, A. I. (2020). Intrudtree: A machine learning based cyber security intrusion detection model. *Symmetry, 12*(5), 754.
28. Liu, H., & Motoda, H. (Eds.). (1998). *Feature extraction, construction and selection: A data mining perspective* (Vol. 453). Springer.
29. Sarker, I. H., Abushark, Y. B., & Khan, A. I. (2020). Contextpca: Predicting context-aware smartphone apps usage based on machine learning techniques. *Symmetry, 12*(4), 499.
30. Han, J., Kamber, M., & Pei, J. (2011). Data mining concepts and techniques third edition. *The Morgan Kaufmann Series in Data Management Systems, 5*(4), 83–124.
31. Witten, I. H., & Frank, E. (2002). Data mining: Practical machine learning tools and techniques with Java implementations. *ACM SIGMOD Record, 31*(1), 76–77.
32. Pedregosa, F., Varoquaux, G., Gramfort, A., Michel, V., Thirion, B., Grisel & Duchesnay, E. (2011). Scikit-learn: Machine learning in Python. *The Journal of Machine Learning Research, 12,* 2825–2830.
33. Pearson, K. (1901). LIII. On lines and planes of closest fit to systems of points in space. *The London, Edinburgh, and Dublin Philosophical Magazine and Journal of Science, 2*(11), 559–572.
34. Hotelling, H. (1933). Analysis of a complex of statistical variables into principal components. *Journal of Educational Psychology, 24*(6), 417.

Chapter 5
Discretization of Time-Series Behavioral Data and Rule Generation based on Temporal Context

5.1 Introduction

In general, discretization is a process to convert continuous numerical attributes into discrete or nominal attributes with a finite number of intervals, resulting in a non-overlapping partition of a continuous domain. As mentioned earlier, the temporal context, represented as time-series data, is the most important aspect that influences user behavior in a mobile Internet portal [1]. Individuals' behaviors vary over time in the real world, and the smart devices record the precise time information (e.g., 2015-04-25 08:35:55) of all the diverse activities with mobile phones in time order, are considered as time-series behavioral data. However, unlike digital systems, human perception of time is not precise. Routine behaviors always have a time interval, even if it is only a small one, such as 5 min [2]. For example, a user calls her mother in the evening regularly. She is unlikely to contact her mother every day at 6:00 p.m.; she could call at 6:13 p.m. one day and 5:51 p.m. on another day. As a result, the exact time is not very informative in predicting user behavior in the future. According to Farrahi et al. [3] time-based effective behavior modeling is an open problem. In this chapter, we explore the discretization process of the continuous time-series data to generate temporal segments according to the behavioral patterns of the users, which is used as the basis of generating rules based on temporal context.

To evaluate time as a condition in a high confidence rule, time must be separated into meaningful categories that act as a proxy for recognizing a user's various activities. Researchers employ numerous methods of segmentation to mine mobile user behavior for diverse goals, including large interval or small interval segmentation without taking into account individual behavioral patterns. For instance, a number of researchers [4–7] use large interval-based segmentation (e.g., morning[6:00 a.m.–12:00 p.m.]) in order to mine mobile user behavior. Such large segments may not suitable for generating meaningful behavioral rules in a temporal context. On Monday, for example, one user may attend a regular meeting from 8:00 to 8:30 a.m., while another attends class from 10:00 to 11:30 a.m. While in a meeting or class, both users reject the incoming call. Users may normally accept incoming calls at different times in the morning. Because of the inability to distinguish individual's such distinct behaviors in the morning, these logged call response behaviors would not be generalizable to a meaningful rule if long static segments (e.g., morning) were used.

On the other hand, several researchers use small interval-based segmentation (e.g., 15 min) [8–10] instead of the above large categories by taking into account the frequent variations of individual's behaviors. However, in many circumstances, these small interval time segments will not yield meaningful rules. If the time interval is too short, for example, there may not be enough behavioral data instances in each segment to determine the dominant behavior based on many observations, or there may be none at all. Creating behavioral rules based on observations with so little "support" is unlikely to be effective [2]. In general, as the period is increased, we should expect more data instances, more support, and more behavioral variations to be detected, masking the true dominating behavior. Such segments are not suited for capturing the actual behaviors of mobile phone users because each individual's activity is different. As a result, to create effective temporal rules, specific behavior-oriented time segments that reflect logged behavior of a single mobile phone user must be discovered.

In this chapter, we present a method for mining an individual's behavior by analyzing their mobile phone time-series data and discovering behavior-oriented time segments. When time is segmented effectively, high confidence rules emerge that capture dominant behavior for as much of the week as possible. To produce rules, association rule learner [11] rather than classification rule learner is considered. The reason is that classification learners cannot ensure that a discovered classification rule will have a high predictive accuracy [12]. Association rule learning, on the other hand, is a well-defined, deterministic task that discovers rule sets with confidence levels over a predetermined threshold [11]. The threshold for developing rules will be established according to an individual's preference for how interventionist the agent should be. As no two people's behaviors are the same in the real world, the segmentation and corresponding rules may differ from user to user based on their behavioral patterns overtime of the week.

5.2 Requirements Analysis

In this section, we summarize the key requirements for an effective time-series segmentation to mine behavioral rules of individual mobile phone users. These are:

Req1 *Dynamic*: There are two main types of segmentation techniques (static and dynamic), in terms of determining the number of segments. As the mobile phone usage behaviors of individuals can be very different, a fixed number of predefined segments is not an effective way to capture individuals' behavior. For instance, only two segments (office hours and non-office hours) may capture the behavioral patterns of one individual, and more number segments (e.g., 24 segments in a day, a 1-h interval for each or more segments) might be needed to capture the behavior of another individual depends on her unique behavioral patterns. Therefore, a static number of arbitrary time segments is not able to capture an individual's behavioral patterns. In addition to the time of the day, an individual's behavior may differ between days of the week (Monday, Tuesday, ..., Sunday). For instance, one's Monday's behaviors are quite consistent and as a result, a fewer number of segments may be able to capture the behavioral patterns, while more segments might be needed for other days of the week for that user. As we have no prior knowledge about an individual's behavior, the segmentation should be dynamic, i.e., the segmentation technique needs to identify the number of optimal segments dynamically without any prior knowledge by analyzing the characteristics of an individual's mobile phone data and eventually extract a set of effective time segments with associated days-of-the-week to produce high confidence temporal behavior rules of individuals.

Req2 *Temporal Coverage Maximization*: Coverage maximization is another requirement while doing an effective segmentation to mine the temporal behavioral rules of individual mobile phone users. The purpose of maximizing temporal coverage is to enrich discoverability by specifying periods that are spanned by similar behavioral patterns of individuals. Simply, temporal coverage refers to how much of the week is covered by the rules. Temporal coverage can be more flexible and determines the corresponding time window (e.g., 10:30 a.m.– 12:15 p.m. is an example of a time window) for which temporal rules must be valid, but this time window may not be the same for all days-of-the-week. Higher temporal coverage makes the temporal rule-set more effective in terms of support value and corresponding applicability while mining mobile phone user's behavior. Therefore, the technique needs to maximize the temporal coverage for each segment as long as the corresponding temporal rule is valid for a particular confidence threshold preferred by an individual user.

Req3 *Accuracy Maximization in Temporal Rules*: The confidence of a rule is directly associated with the accuracy, i.e., a higher value of confidence ensures higher accuracy of rules and vice-versa. To evaluate time as a condition in a high confidence rule, time must be segmented into meaningful categories that serve as a proxy for identifying the user's diverse behaviors in her daily life activities. For instance, to get a high confidence value, if we take into account a very small segment assuming like behavior in that segment, the temporal rule with that segment may have very low support that may not be meaningful. In general, by increasing the time interval we would expect more data instances (greater support) but also greater behavioral variations that reduce the confidence of the rule. As mentioned above, the mobile user behavior may not be consistent at various time-of-the-day and days-of-the-week, the technique needs to generate optimal time segments according to their similar behavioral patterns, to produce effective temporal rules that satisfy the confidence threshold (accuracy label) preferred by an individual mobile phone user.

In this work, we address the above three aspects for generating behavior-oriented time segments capturing similar behavior characteristics according to individual's unique behavioral patterns as the basis for mining their temporal behavior rules.

5.3 Time-series Segmentation Approach

In this section, we discuss how does the behavior-oriented time segmentation (BOTS) technique work to extract temporal behavior rules from time-series mobile phone data to mine people's behavior.

5.3.1 Approach Overview

In this approach, first, a small base period is taken into account to divide each day of the week into relatively little temporal slices. For this, we consider a 5-min interval to be the finest granularity required to distinguish an individual user's day-to-day activity. Then behavior-oriented segments are generated from the time slices generated by a specific base period mentioned above. To do so, we determine each slice's dominant behavior and dynamically combine adjacent slices with similar dominating characteristics to produce larger segments of similar behavior. These aggregated segments will provide greater support and coverage over time, and they can be utilized as the foundation for mining rules specific to individuals. Finally, we choose the best segmentation by determining the segmentation's applicability. We iteratively increase the basis period and compare the applicability of the corresponding segmentation over each iteration to discover the optimal base period because we have no prior knowledge of individual behavioral patterns. The optimal time segmentation is determined by the time segmentation that gives the greatest applicability, and the related base period is used as the optimal base period for capturing individual behavioral patterns. Finally, we generate the temporal behavior rules for the users based on the discovered optimal segmentation. The block diagram of this process for obtaining temporal behavior rules of individuals is shown in Fig. 5.1. The components of the above figure are described one by one in the subsections that follow.

5.3.2 Initial Time Slices Generation

As the approach is based on individual behavior, the first part of this method is the generation of initial time slices for collecting an individual's behavioral activity throughout 24 h. To accomplish this, we start by dividing each day of the week into small temporal slices based on a base period. These initial time slices are utilized to

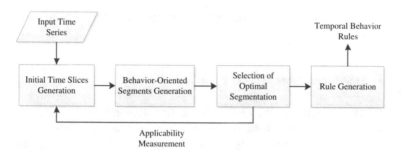

Fig. 5.1 An overview of time-series segmentation approach for mining temporal behavior rules of individual mobile phone users

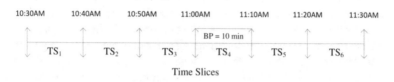

Fig. 5.2 An example of initial time slices for a particular base period

capture their behavioral patterns, as people's daily behavior occurs in a period rather than at a specific time. The number of time slices is determined by the base period's duration. If T_{max} represents the entire 24-h day and BP represents the base period, the number of slices is

$$Number\text{-}of\text{-}Slices = \frac{T_{max}}{BP} \tag{5.1}$$

The number of time slices reduces as the base period increases, according to Eq. (5.1). If the starting base period is 5 min, the number of slices will be $(24\text{-}hours\text{-}a\text{-}day)/5 = 288$. A base period, such as 5 min, is considered to be the finest granularity for distinguishing an individual's day-to-day activities. The number of slices will be $(24\text{-}hours\text{-}a\text{-}day)/10 = 144$ if the base period is increased to $(5 times 2) = 10$ min in the second iteration. Figure 5.2 illustrates an example of initial time slices $(T S_1, ..., T S_6)$ with time limits between 10:30 a.m. and 11:30 a.m., when the base period (BP) is 10 min.

5.3.3 Behavior-Oriented Segments Generation

In this section, we generate the behavior-oriented time segments based on the initial time slices generated for a particular base period. This section divides into two parts: the first one is dominant behavior identification for the generated time slices, and the second one is their dynamic aggregation based on the identified similar dominant characteristics.

5.3.3.1   Dominant Behavior Identification

As we take into account the diverse behavioral activities of individuals across time, we first identify the dominant behavior for each time slice generated in the previous phase. The largest number of occurrences of a given activity among a list of activities in a time slice is represented as dominant behavior [2]. We divide the activity instances from the log into time slices since an individual's behavior pattern differs depending on the duration of the typical activities they do during the week. We take the entire period to be a week in this case, assuming that individuals' typical behaviors follow a weekly pattern. As a result, activities from many weeks for the same weekly time slices are merged, and the entire week is divided into time slices.

The time slice containing the dominant behavior may play a role in generating a high confidence rule for that dominating behavior. We may not receive dominating behavior in some time slices since we have no prior knowledge about an individual's behaviors over time of the week.

Assume we have a time slice TS_{30} with the following behavioral information, where the first parameter reflects the corresponding occurrences in percentage in TS_{30} and the second parameter reflects user behavior class.

$$\{TS_{30} : (BH_1, 45\%), (BH_2, 45\%), (BH_3, 10\%)\}$$

However, because both BH_1 and BH_2 have the same number of occurrences, there is no dominating behavior in TS_{30} (45%). As a result, using TS_{30} to generate rules results in numerous rules with conflict behaviors (BH_1 and BH_2), which is impractical. We can avoid such conflicting rules in terms of rule confidence by taking into account more than 50% of occurrences for specific behavior in a time slice.

Assume we have another time slice TS_{35} with the behavioral information below, where the parameters reflect user behavior class and associated occurrences in percentage in TS_{35}, respectively.

$$\{TS_{35} : (BH_1, 55\%), (BH_2, 40\%), (BH_3, 5\%)\}$$

As a result, BH_1 is the dominating behavior in TS_{35}, with the highest occurrences (55%) compared to the others. The time slice TS_{35} may play a role in generating a relevant conflict-free rule with the dominating behavior BH_1. However, as previously said, the confidence level for creating rules varies depending on an individuals' preferences for how interventionist they want to be. Consider the case where a user U_x's chosen confidence threshold is 80%, implying that she is not interested in rules with the confidence of less than 80%. Even though there is a clear dominating behavior (BH_1) in that time slice, the produced rule utilizing the time slice TS_{35} will be useless for U_x.

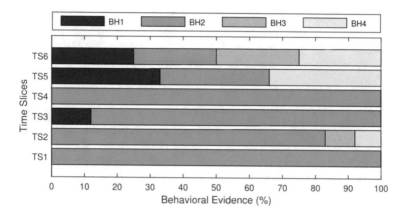

Fig. 5.3 Sample behavioral data (%) in different time slices

As a result, we use the preferred rule confidence threshold (t) to determine the dominant behavior of each time slice to construct behavioral rules based on individual preferences. Using this threshold has the advantage of reducing the amount of processing required to obtain the desired segmentation based on individual preferences. If the proportion of a specific behavior class $BH \geq threshold(t)$ in a time slice, then BH_i is the dominant behavior for that time slice. Figure 5.3 presents a sample behavioral data evidence for recognizing dominant behavior for various time slices, assuming a confidence threshold of 80%.

Figure 5.3 shows that TS_1 has 100% BH_2 that meets the confidence, indicating that BH_2 is the dominant behavior for this slice. BH_2 is the dominant behavior for the TS_2 slice because it also meets the threshold as TS_2 has 83% BH_2, 8% BH_3, and 9% BH_4. Similarly, for time slices TS_3 and TS_4, BH_2 is the dominant behavior. There is no dominating behavior for the time slices TS_5 and TS_6 as no behavior larger than 80% is obtained in these two slices. Overall, each slice can only have one dominant behavior as the dominant behavior is the one with the greatest number of occurrences, according to the definition. If TS_{total} is the total number of time slices, then the number of time slices containing the dominant is

$$Number\text{-}of\text{-}TS(dominant) \leq TS_{total}$$

5.3.3.2   Dynamic Aggregation

Once the dominant behavior for each time slice has been discovered, slices that exhibit the same dominant behavior are dynamically aggregated into the longest possible time segments using our methodology. This is done to increase the support value and temporal coverage of any rules that are retrieved for these time segments later on.

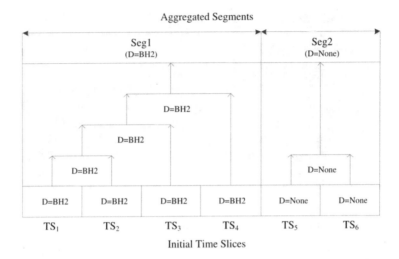

Fig. 5.4 Dynamic aggregation of initial time slices based on the dominant behavior

Assume we have four consecutive time slices TS_1, TS_2, TS_3, and TS_4, each with the behavioral information shown in Fig. 5.3, where the first parameter represents the time slice and the second parameter denotes the corresponding dominant behavior for that time slice.

$$\{(TS_1, BH_2), (TS_2, BH_2), (TS_3, BH_2), (TS_4, BH_2)\}$$

Because each of these time slices has the dominating behavior, each slice can create significant rules in terms of confidence on its own. However, because these time slices include the same dominating behavior, we aggregate them into one single longest segment Seg_1 (shown in Fig. 5.4) to achieve an effective behavior-oriented segment. As a result, the dominating behavior's longest equivalent behavioral segment can yield more relevant rules in terms of support, temporal coverage, and confidence with the dominant behavior BH_2.

We employ a bottom-up hierarchical aggregation strategy based on dominating behavior to find the longest related behavioral segments. The agglomerative clustering algorithm [13], which uses a proximity matrix built by computing the distance between clusters, is the most similar methodology. The method merges the clusters one by one based on the matrix value until the desired cluster structure is achieved, which is set by a threshold. Because of the differences in users' behavior over time, predicting the threshold level at which the merging works best according to a proximity matrix is quite challenging. As a result, we create subsequent segments by dynamically aggregating beginning time slices based on similar behavior, with some segments requiring more merging and others requiring less merging, as the behavior of individual mobile users varies. Figure 5.4 provides a sample example of employing dynamic aggregation to create such dynamic segments $[Seg_1, Seg_2]$ from the original time slices, where BH_2 is the dominant behavior of Seg_1 $[D = BH_2]$ and Seg_2 $[D = None]$ has no dominant behavior.

Algorithm 1: Dynamic aggregation

Data: initial time slices list: TS_{list}
Result: behavior-oriented segment list: Seg_{list}

1 //create initial segment using the first time slice
2 $Seg_{init} \leftarrow TS_1$
3 //insert segment into the segment list
4 $Seg_{list} \leftarrow insert(Seg_{init})$
5 **foreach** TS in TS_{list} **do**
6   //identify dominant behavior using the threshold t
7      $D \leftarrow identifyDominant(TS, t)$
8   //check the dominant behavior
9   **if** $D(Seg_{init}) \equiv D(TS)$ **then**
10   //aggregate into one segment
11         $Seg_{agg} \leftarrow aggregate(Seg_{init}, TS)$
12   //initial segment is changed to aggregated segment
13         $Seg_{init} \leftarrow Seg_{agg}$
14   //update segment list
15         $Seg_{list} \leftarrow update(Seg_{init})$
16   **else**
17   //create new segment using the next time slice
18         $Seg_{new} \leftarrow createSeg(TS)$
19   //insert segment into the list
20         $Seg_{list} \leftarrow insert(Seg_{new})$
21   **end**
22 **end**
23 return Seg_{list}

Algorithm 1 explains how to perform this dynamic aggregation. The list of starting time slices TS_{list} is used as input data, and the list of behavior-oriented segments Seg_{list} is used as output data. Our approach Algorithm 1 dynamically determines the number of segments to be formed from an individual's time-series mobile phone data, rather than arbitrarily determining the number of segments in advance. As a result, depending on their behavioral patterns, the number of segments and time duration of the produced segments will vary from user to user.

5.3.4   Selection of Optimal Segmentation

In this section, we find the best segmentation and behavior-oriented segments that may be utilized to construct individual mobile phone users' temporal behavior rules.

5.3.4.1   Segments Filtering

As different lengths of segments with different dominant behaviors (for example, Seg_1 with $D = BH_2$ and Seg_2 with $[D = None]$, as shown in Fig. 5.4) are produced after dynamic aggregation, we must choose segments that can produce high confidence temporal rules to reduce the processing burden. The reason for this is that employing all of the segments generated by dynamic aggregation is unlikely to yield high confidence behavioral rules since all of these segments may not reflect the strong pattern of a particular behavior that is required to represent a segment as a rule.

We simply ignore segments that have no dominant behavior (e.g., segments with $[D = None]$) when selecting segments that can construct behavioral rules based on the desired confidence of individuals. Because of employing segments with $[D = None]$, it is impossible to build temporal rules that satisfy the user's preferred confidence. As a result, to produce meaningful temporal behavior rules for individuals, we only take into account segments that exhibit a specific dominant behavior.

Assume we have three segments with the behavioral data below, where the first parameter specifies time segments and the second parameter signifies the matching dominating behavior after dynamic aggregation.

$$\{(Seg_1, BH_2), (Seg_2, None), (Seg_3, BH4)\}$$

Since Seg_2 has no dominant behavior, it is unable to generate any meaningful behavioral rules based on the individual's preferences. As a result, we minimize the segment size by filtering such segments and consider Seg_1 and Seg_3 for discovering rules, as each of these segments contains distinct dominating behavior that serves as the foundation of useful behavioral rules of the users.

5.3.4.2   Applicability Measurement

Because of their effects on support, temporal coverage, and confidence, different base periods may result in various time segmentation and related rules. We take into account each of the filtered segments with the dominating behavior is an antecedent of the temporal rule for measuring applicability because they are all capable of producing rules according to individual preferences.

We present a metric called 'applicability' that measures the applicability of rules generated by the aforementioned filtered segments with a certain dominating behavior to find the best segmentation. Applicability is a descriptive statistic that considers two parameters for a certain confidence level. These are:

- Temporal coverage—It is the time interval that a temporal rule covers. If $tstart$ and $tend$ are the start and end time points of a particular time segment that is used to generate a temporal rule R, then the temporal coverage for that rule $Rcov = |tend - tstart|$, i.e., the internal time interval of that segment.

- Support—The number of behavioral instances in a time segment used to generate a temporal rule is known as support (Rsup).

Time segmentation during the week is used as a proxy for the user's activities and subsequent behavior in our method. On the one hand, we want time segmented with sufficient resolution to distinguish between different forms of dominant behavior for a given confidence threshold. We also want as much support as feasible for rules that capture that behavior. However, the association rule learning metrics of confidence and support [11] are insufficient for identifying appropriate temporal rules for mining mobile user behavior. The reason for this is that temporal rules might have a little or big temporal coverage, depending on the volatility of a user's behavior stability across time. The traditional metric treats each context (for example, a time segment with a small or long time interval) as a distinct item that is more significant in market basket analysis [11]. As a result, it does not capture the role of temporal coverage in identifying relevant user behavioral rules.

We define our new 'applicability' *metric* as follows:

Applicability: It is defined as the product of aggregate support and aggregate temporal coverage, where aggregate support is the fraction of the sum of all the rules' support counts that satisfies the confidence threshold among the maximum possible support considered, and aggregate temporal coverage is the proportion of the temporal coverage by those rules.

Formally, the applicability is defined as:

$$Applicability = \sum_{i=1}^{N'} \left(\frac{Rsup_i}{S_{max}} * \frac{Rcov_i}{C_{max}} \right) \tag{5.2}$$

where $Rsup$ represents a rule's support count, $Rcov$ represents the rule's temporal coverage, S_{max} represents the maximum possible support in a dataset, C_{max} represents the maximum possible temporal coverage in a week, and N' represents the number of rules that satisfy the user's confidence threshold.

5.3.4.3 Identify Optimal Segmentation

As previously stated, the applicability of temporal rules for a certain confidence threshold is determined by the dynamic segments list generated, which is based on the length of the base period. The best segmentation will be determined by the unique pattern of the user's various behaviors. We iteratively increase the base period by a suitable time gap and compare the applicability of the associated segmentation over each iteration to discover the optimal base period because we have no prior information of individual behavioral patterns. The optimal time segmentation is determined by the generated time segments that give the maximum applicability, and the optimal base period is determined by the corresponding base period that captures the unique behavioral patterns of individuals. As our method is based on personalized behavioral activities, the optimal base period for capturing the behavioral pattern and the corresponding segments for developing temporal behavior rules may differ from one user to another in our real-world life.

Algorithm 2: Identify optimal segmentation

Data: base period: BP
Result: optimal segments list: $OSeg_{list}$

1 //initialize applicability
2 $A_{init} \leftarrow 0$
3 **foreach** BP *in 24-hours-a-day time scale* **do**
4 $\quad$ //generate initial time slices using base period
5 $\quad$ $TS_{list} \leftarrow$ generateTS(BP)
6 $\quad$ //produce behavior-oriented aggregated segments
7 $\quad$ $Seg_{list} \leftarrow$ aggregateSeg(TS_{list})
8 $\quad$ //get filtered segments
9 $\quad$ $FSeg_{list} \leftarrow$ filterSeg(Seg_{list})
10 $\quad$ //calculate the applicability utilizing filtered segments
11 $\quad$ $Applicability \leftarrow$ calculateApplicability($FSeg_{list}$)
12 $\quad$ //compare the applicability
13 $\quad$ **if** $Applicability > A_{init}$ **then**
14 $\quad\quad$ //store the base period as optimal base period
15 $\quad\quad$ $BP_{optimal} \leftarrow BP$
16 $\quad\quad$ //update initial applicability
17 $\quad\quad$ $A_{init} \leftarrow Applicability$
18 $\quad\quad$ //update optimal list
19 $\quad\quad$ $OSeg_{list} \leftarrow$ updateOSegList(Seg_{list})
20 $\quad$ **end**
21 $\quad$ //next base period
22 $\quad$ increase BP
23 **end**
24 return $OSeg_{list}$

Algorithm 2 depicts the overall process. The base period BP is used as input data, while the list of optimal segments $OSeg_{list}$ is used as output data.

5.3.5 Temporal Behavior Rule Generation Using Time Segments

We apply the well-known association rule learning algorithm Apriori [11] to generate the temporal rules of an individual user using the best segmentation. We chose this technique because we wish to build a set of temporal rules using the segments generated above. One of the most important advantages of employing association rule learning is that a discovered behavioral rule will have a high prediction accuracy [12] because it allows an individual to create rules based on her preferred level of confidence.

Although this approach has the disadvantage of redundancy in associations, it does not produce redundant rules in our situation since we only employ the temporal segments to generate the output temporal rules. Furthermore, both end-users and application developers will find it simple to read and understand [14].

A temporal rule is expressed as *temporal context* $\Rightarrow$ *behavior*, with the antecedent being "temporal context" and the consequent being "user behavior." The algorithm constructs rules with an antecedent that contains temporal information [weekday, time segment] and a consequent that solely contains individual behavior during that time period. This means that rules can be in the form *temporal context* $\Rightarrow$ *behavior* but not in the form of *behavior* $\Rightarrow$ *temporal context*. To better understand the concept of temporal rules let us consider an example of phone call behaviors where the user: (i) always (100%) makes outgoing calls between 13:00 and 14:00 on Thursdays; (ii) rejects most of the incoming calls (92%) between 14:10 and 15:45 on Mondays; (iii) misses most of the incoming calls (87%) between 19:00 and 21:00 on Saturdays, and (iii) accepts most of the incoming calls (98%) between 07:30 and 10:30 on Sundays, then the following temporal rules would represent the user's behavior for these temporal segments:

(i) $Thursday$ $[13:00-14:00]$ $\Rightarrow$ $Outgoing$ $(Conf = 100\%)$

(ii) $Monday$ $[14:10-15:45]$ $\Rightarrow$ $Reject$ $(Conf = 92\%)$

(iii) $Saturday$ $[19:00-21:00]$ $\Rightarrow$ $Missed$ $(Conf = 87\%)$

(iv) $Sunday$ $[07:30-10:30]$ $\Rightarrow$ $Accept$ $(Conf = 98\%)$

By validating the parameters 'support' and 'confidence' defined above, the algorithm examines the data and generates such temporal rules. In other words, a set of temporal behavioral rules for an individual mobile phone user is developed only when it satisfies the user's specified minimal support and confidence criteria. It's worth mentioning that lowering the support or confidence levels can lead to the discovery of more temporal rules, and vice versa [11].

5.4 Effectiveness Comparison

In this experiment, we compare the effectiveness of our BOTS methodology to existing time segmentation methodologies in terms of applicability and data coverage (percentage). To accomplish so, we first choose five baseline methods for mining mobile user behavior that use distinct time segments. For comparison purposes, we denote these baseline methods as BM1 [8] that uses 15-min equal interval for time segmentation to mine human mobility patterns, BM2 [4] that uses 4-unequal time slots based segmentation for learning mobile user preferences for notification management, BM3 [15] that uses 5-unequal time slots for time segmentation for mining mobile user preferences for personalized recommendation, BM4 [7] that uses 4-h equal interval-based time segmentation for learning phone usages sequential patterns to build mobile sequence mining engine and finally, BM5 [9] that uses 3-h equal interval for time segmentation to identify human daily activity patterns utilizing mobile phone data respectively. To compare the techniques objectively, we averaged behaviors from different weeks using the same datasets for various baseline techniques.

To show the effectiveness for individual users, Figs. 5.5 and 5.6 show the relative comparison of applicability and Figs. 5.7 and 5.8 show the relative comparison of data coverage (%) for User X and Y respectively. As no rule is meaningful without the minimum number of occurrences, we take into account minimum support 1 (one instance) for each approach in our experimental purpose. Furthermore, we investigated several confidence thresholds, including 51% (lowest strength), 60%, and up to 100% (maximum strength).

From Figs. 5.5, 5.6, 5.7, and 5.8, we find that our BOTS approach consistently outperforms previous approaches for different confidence thresholds. The fundamental reason is that existing approaches to mining mobile user activity do not take into account people's various behavioral patterns for segmentation. Our dynamic technique, on the other hand, is behavior-oriented and can better capture the unique behavioral patterns of each user, resulting in a set of behavior-oriented segments for a certain confidence threshold.

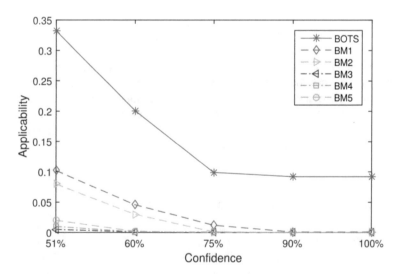

Fig. 5.5 Applicability comparison of different segmentation approaches utilizing an individual's mobile phone data (User X)

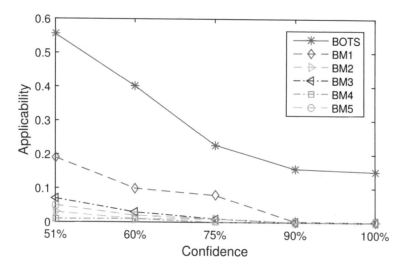

Fig. 5.6 Applicability comparison of different segmentation approaches utilizing an individual's mobile phone data (User Y)

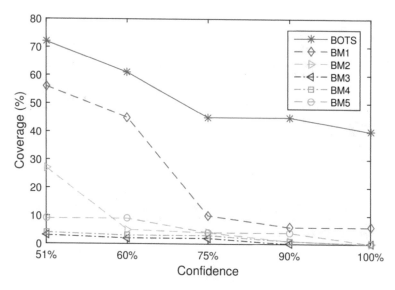

Fig. 5.7 Data coverage (%) comparison of different segmentation approaches utilizing an individual's mobile phone data (User X)

In addition to individual comparison, we also show the relative comparison of average applicability and data coverage (%) for a collection of users of two different datasets shown in Fig. 5.9. For this, we calculate the average applicability and data coverage (%) of all the users taken for experimental purposes in this study, for each approach with the same confidence threshold 75%. The average findings also reveal

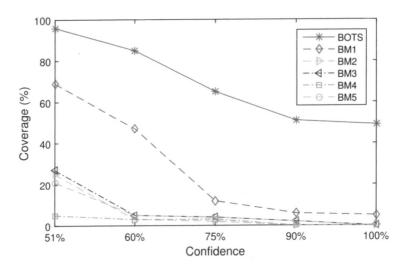

Fig. 5.8 Data coverage (%) comparison of different segmentation approaches utilizing an individual's mobile phone data (User Y)

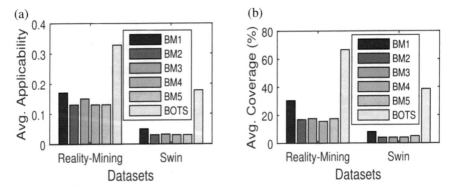

Fig. 5.9 Average applicability and average coverage comparison of different segmentation approaches utilizing the collection of individuals mobile phone data of different datasets. (**a**) Average applicability. (**b**) Average coverage

that for a group of users, our BOTS methodology continuously outperforms earlier approaches. The reason is that we can more accurately identify each user's unique behavioral patterns, resulting in higher applicability and data coverage (percentage) values for all users. However, existing techniques are not behavior-oriented and are unable to represent the user's different behavioral patterns that change over time. As a result, the likelihood of hiding the real dominant behavior in a segment rises in tandem with other existing behaviors, lowering the applicability and data coverage (percentage) for a given confidence level. By better capturing individual's behavioral patterns, our dynamic time segmentation methodology overcomes these issues and enhances segmentation quality in terms of applicability and data coverage (percent) for a given confidence threshold.

5.5 Conclusion

We have presented a dynamic behavior-oriented time segmentation methodology in this chapter for extracting temporal behavior rules to mine mobile user behavior using their phone data. Our method dynamically determines the optimal continuous-time segments, each of which is dominated by the user's activity. As a result, for these time segments, temporal rules are generated, which can be used to construct an automated rule-based personal assistance system for mobile phone users. The time segments are determined based on the users' contiguous dominant activity, can span the week, and will differ from user to user to accurately reflect their behavioral patterns. Furthermore, the time segments and behavioral rules are chosen in such a way that the preferred confidence threshold achieves maximum temporal coverage by the rules, resulting in maximum applicability for the rules. We have also established the applicability measure for this reason, which considers the support and temporal coverage that the mined rules provide. Although we use phone call behavior scenarios as examples throughout the chapter, our method is also applicable to different application domains, e.g., in the context of cybersecurity, this method can help for effectively analyzing the behavioral patterns of various cyber-attacks in time series.

References

1. Halvey, M., Keane, M. T., & Smyth, B. (2006). Time based patterns in mobile-internet surfing. In *Proceedings of the SIGCHI Conference on Human Factors in Computing Systems* (pp. 31–34).
2. Sarker, I. H., Colman, A., Kabir, M. A., & Han, J. (2018). Individualized time-series segmentation for mining mobile phone user behavior. *The Computer Journal, 61*(3), 349–368.
3. Farrahi, K., & Gatica-Perez, D. (2014). A probabilistic approach to mining mobile phone data sequences. *Personal and Ubiquitous Computing, 18*(1), 223–238.
4. Mehrotra, A., Hendley, R., & Musolesi, M. (2016). PrefMiner: Mining user's preferences for intelligent mobile notification management. In *Proceedings of the 2016 ACM International Joint Conference on Pervasive and Ubiquitous Computing* (pp. 1223–1234).
5. Song, Y., Ma, H., Wang, H., & Wang, K. (2013). Exploring and exploiting user search behavior on mobile and tablet devices to improve search relevance. In *Proceedings of the 22nd international conference on World Wide Web* (pp. 1201–1212).
6. Xu, Y., Lin, M., Lu, H., Cardone, G., Lane, N., Chen, Z., ... & Choudhury, T. (2013). Preference, context and communities: a multi-faceted approach to predicting smartphone app usage patterns. In *Proceedings of the 2013 International Symposium on Wearable Computers* (pp. 69–76).
7. Mukherji, A., Srinivasan, V., & Welbourne, E. (2014). Adding intelligence to your mobile device via on-device sequential pattern mining. In *Proceedings of the 2014 ACM International Joint Conference on Pervasive and Ubiquitous Computing: Adjunct Publication* (pp. 1005–1014).

8. Ozer, M., Keles, I., Toroslu, H., Karagoz, P., & Davulcu, H. (2016). Predicting the location and time of mobile phone users by using sequential pattern mining techniques. *The Computer Journal, 59*(6), 908–922.

9. Phithakkitnukoon, S., Horanont, T., Di Lorenzo, G., Shibasaki, R., & Ratti, C. (2010, August). Activity-aware map: Identifying human daily activity pattern using mobile phone data. In *International workshop on human behavior understanding* (pp. 14–25). Berlin, Heidelberg: Springer.

10. Do, T. M. T., & Gatica-Perez, D. (2014). Where and what: Using smartphones to predict next locations and applications in daily life. *Pervasive and Mobile Computing, 12*, 79–91.

11. Agarwal, R., & Srikant, R. (1994). Fast algorithms for mining association rules. In *Proc. of the 20th VLDB Conference* (Vol. 487, p. 499).

12. Freitas, A. A. (2000). Understanding the crucial differences between classification and discovery of association rules: a position paper. *ACM SIGKDD Explorations Newsletter, 2*(1), 65–69.

13. Xu, R., & Wunsch, D. (2005). Survey of clustering algorithms. *IEEE Transactions on Neural Networks, 16*(3), 645–678.

14. Srinivasan, V., Moghaddam, S., Mukherji, A., Rachuri, K. K., Xu, C., & Tapia, E. M. (2014). Mobileminer: Mining your frequent patterns on your phone. In *Proceedings of the 2014 ACM International Joint Conference on Pervasive and Ubiquitous Computing* (pp. 389–400).

15. Zhu, H., Chen, E., Xiong, H., Yu, K., Cao, H., & Tian, J. (2014). Mining mobile user preferences for personalized context-aware recommendation. *ACM Transactions on Intelligent Systems and Technology, 5*(4), 1–27.

Chapter 6
Discovering User Behavioral Rules Based on Multi-Dimensional Contexts

6.1 Introduction

Real-world smartphone data usually comprise a set of features relevant to the individual user activities, whose interpretation depends on some *contextual information* such as temporal context (day, time), spatial context (e.g., location - at office), social context (social activity, e.g., meeting, or social relationship between individuals, e.g., mother). In this chapter, we take into account these multi-dimensional contexts for discovering user behavioral rules. As defined in earlier chapter, a rule ($A \Rightarrow C$) is any statement that relates two principal components, the rule's left-hand side (antecedent, A) and the rule's right-hand side (consequent, C) together. An antecedent states the condition (IF) and a consequent states the result (THEN) held from the realization of this condition, i.e., (IF-THEN logical statement). According to the general definition of a rule, we define a behavioral rule of an individual mobile phone user based on multi-dimensional contexts, as [*contexts* $\Rightarrow$ *user behavior*], where *contexts* (antecedent) represents the information of such multi-dimensional contexts, and *behavior* (consequent) represents individual's mobile phone usage behavior for that contexts. An example of such a behavioral rule of an individual mobile phone user would be "On Monday[10:00–12:00] (temporal), if a user is in a meeting (social activity) at her office (locational), she rejects (behavior) the incoming calls of her mobile phone" and represented as a rule format as "*Monday*[10:00–12:00], *Meeting*, *Office* $\Rightarrow$ *Reject*". A set of discovered such behavioral rules of individual mobile phone users based on multi-dimensional contexts utilizing their mobile phone data can be used to provide personalized services to intelligently assist them in their daily activities in a context-aware pervasive computing environment.

I. H. Sarker et al., *Context-Aware Machine Learning and Mobile Data Analytics*,
https://doi.org/10.1007/978-3-030-88530-4_6

In the area of mining mobile phone data, association rule mining [1] and classification rule mining [2] are the most common techniques to discover such types of rules of individual mobile phone users. In particular, several researchers [3–5] have used classification rule mining technique (e.g., C4.5 Decision tree)[2] to mine rules capturing mobile phone users' behavior. However, the decision tree-based rules mostly have low reliability [6, 7]. The reason is that in some cases, the performance of decision tree-based rules is very low because of having lower confidence value and consequently gives lower prediction accuracy for unseen test cases. Due to the over-fitting problem and inductive bias, decision trees cannot ensure that a discovered classification rule will have a high predictive accuracy [8]. Furthermore, this methodology lacks the flexibility to specify user preferences (e.g., confidence level) that may differ from user to user based on the consistency of their behaviors, resulting in mobile phone users making rigid decisions [2].

On the other hand, the association rule mining technique (e.g., Apriori) [1] is well defined in terms of the rule's performance (e.g., accuracy) and flexibility as it has the own parameter support and confidence [8]. Several researchers [6, 9, 10] have used the association rule mining technique (e.g., Apriori)[1] to mine rules capturing mobile phone users' behavior. The association rule mining methodology finds all context associations in the dataset that meet the user-specified minimum support and confidence preferences. As a result of examining all possible combinations of contexts without any intelligence, it generates a large number of redundant rules. For instance, if a particular context (e.g., meeting) can take a call handling decision of an individual user, then the combination of additional contexts (e.g., meeting, office, Monday) for the same decision making, is considered a redundant rule for that user. Such redundant discovery affects the quality and usefulness of the rules. According to Fournier et al. [11], the association rule mining technique generates up to 83% redundant rules, resulting in an unnecessarily large ruleset, which makes the decision-making process ineffective and more complex [12]. Furthermore, it is not feasible to provide real-time, proactive, and personalized services as it requires a significant amount of training time. For instance, Srinivasan et al. [9] observe a long-running time exceeding several hours, when the traditional association rule mining algorithm Apriori [1] is applied to mobile context data.

The above discussion highlights that the traditional association rule mining technique (e.g., Apriori) [1] and the classification rule mining technique (e.g., Decision tree) [2] might not be suitable for mining the useful behavioral rules of individual mobile phone users based on multi-dimensional contexts. In this chapter, we present a tree-based rule mining approach that produces a concise set of useful behavioral rules based on multi-dimensional contexts. This approach generates not only the general behavioral rules of individual mobile phone users but also their specific exceptions while mining rules from the datasets. This approach also provides flexibility in terms of allowing to configure the confidence preference for each mobile phone user while mining their behavioral rules, which can play a significant role in building context-aware personalized applications for the users.

6.2 Multi-Dimensional Contexts in User Behavioral Rules

As we aim to discover user behavioral rules based on multi-dimensional contexts, in this section, we discuss several relevant contexts that might have an influence on individual mobile phone users and can be used in mining their useful behavioral rules. These are:

- *Temporal Context*—In the real world, each user activity is connected with a specific timestamp (e.g., 2015-04-25 08:35:55), hence temporal context is important when modeling people's mobile phone usage behavior. Through the examination of a large sample of user data, Halvey et al. [13] have shown that time-of-week is an essential element in modeling mobile user behavior. This temporal context is related to a specific date (e.g., YYYY-MM-DD), day-of-week (Monday, Tuesday, ..., Sunday), and time-of-day (e.g., hh:mm: ss). Additional temporal information, such as public holidays, weekdays, and weekends, may impact mobile phone users' decisions in addition to the basic temporal information. Humans, unlike digital systems, do not have a precise understanding of time [14]. Routine behaviors always have a temporal interval. As a result, to employ such time-series temporal information in behavioral rules, a time-series discretization approach (e.g., time segments) is required, which is discussed in the previous chapter. Friday [09:00–11:00], Monday [12:00– 13:00], Saturday [15:30–18:45], and so on are some examples of generated time segments.
- *Spatial Context*—Another important user context that may be used to better characterize individual's mobile phone usage behavior is spatial context, e.g., user location [15]. Understanding human mobility in everyday life is essential for applications that provide location-based services. Many emerging location-based applications have been accepted by mobile phone users as the pervasiveness of smart mobile phones has increased in recent years. Some of the factors that contribute to the popularity of locational context in mobile phone applications are highlighted below. First, location-based services rely on the user's geographic location to acquire relevant information on the spot, and thus the user reacts appropriately in that context [16]. For example, a person's phone call response behavior at office may differ significantly from her answer at home. Xu Sun [17] also demonstrated how location data can be used to enhance the mobile user experience by offering relevant mobile services. Second, most modern smartphones can locate or approximate their actual location using various embedded sensors or technologies such as GPS, Wi-Fi, Bluetooth, and so on [16]. In addition to these methods, cell tower ID [18, 19] can be used to determine nominal values of approximate location. Office, home, market, store, restaurant, car, and other coarse level places [20] can be utilized to extract the behavioral rules of individual mobile phone users.

- *Social Context*—In addition to the temporal and geographic settings mentioned above, social factors have an impact on individual mobile phone users' decision-making [21]. Individuals participate in a variety of social activities in the real world, such as professional meetings, seminars, and lectures. People differ significantly in their use of mobile phones during diverse occasions, as demonstrated by Sarker et al. [22]. For example, during an event meeting, one person may be happy to answer incoming phone calls, while another may not [23]. Even a single person may act differently depending on the type of occurrence that has occurred [24]. A person's phone call reaction at a "professional meeting" may differ significantly from her reaction during a "lunch break" event. Individual mobile phone users' calendar schedules can provide such contextual information or factors regarding various types of occurrences [25]. Calendars (e.g., Google Calendar, Outlook Calendar) are commonly used to organize and manage daily activities and schedules. [22, 26]. Individual mobile phone users often use electronic calendars as a personal activity management system to coordinate their daily activities or agendas. These calendars can also be a useful resource for them because they provide various contextual information about the individual's scheduled events or appointments. These can be work-related activities (e.g., professional meetings, classes), family activities (e.g., holidays with family members, picking up kids from school, medical appointments), or public events (e.g., concerts, sporting events, etc.) [27–29]. In addition to such social events, individual social interactions, such as those with family, friends, colleagues, and significant ones, have an impact on individual mobile phone users' decision-making [30, 31]. During an event formal meeting, for example, a user normally rejects an incoming phone call; nevertheless, if the incoming contact is from her boss, she answers. As a result, the social relationship between the caller and the callee has a significant impact on how the call is handled. Thus, social contextual information, in addition to temporal and locational context, can be used to find the useful behavioral rules of individual mobile phone users.

The above discussed multi-dimensional contexts (temporal, locational, and social) are taken into account in this study to mine relevant behavioral rules of individual mobile phone users. In our technique, however, we do not rely on the static number of contexts to mine their behavioral rules. We primarily concentrate on the rule discovery methodology, which is based on a variety of user-relevant scenarios. As a result, we leverage the relevant user's contextual information available in the gathered mobile phone datasets to demonstrate the efficacy of the appropriate behavioral rules established by our technique for individual mobile phone users.

6.3 Requirements Analysis

In this section, we highlight the key requirements for discovering the *useful behavioral rules* of individual mobile phone users based on the multi-dimensional contexts (discussed above) utilizing their mobile phone data. These are:

Req1 *Flexibility*: Flexibility is one of the vital requirements for mining useful behavioral rules of individual mobile phone users. This allows individual mobile phone users to configure their preferences (confidence level) for discovering rules according to their behavioral patterns. In the real world, individual user's behavioral patterns are not identical to all. Moreover, we do not expect 100% like behavior of an individual for a particular context. For instance, a user typically rejects most of the incoming phone calls and accepts a few significant calls such as calls from her boss, when she is in a meeting. In a similar context meeting (*social situation* → *meeting*) the user behaves differently (reject or accept) with the incoming phone calls, which affects the confidence value of rules. If we want to produce rules based on rigid decision-making, we may lose some useful rules that might be interested in a particular user. Thus, unlike a traditional decision tree [2], a behavioral rule mining approach should have *flexibility* for allowing individual mobile phone users to set the confidence level according to their preferences while mining their behavioral rules rather than *rigid decision making*. Such preferences may vary from user to user as the behavioral patterns are not identical to all in the real world. An example of such flexibility is, say, an individual wants the agent to reject calls where in the past he/she has rejected calls more than, say, 80% of the time (e.g., confidence level = 80%). So the rules that satisfy this threshold are discovered for this user. Another individual may want the agent to reject calls where in the past he/she has rejected calls more than, say, 95% of the time. The discovered rules that satisfy this user-specific threshold, are considered as *reliable* rules (good performing) for that user.

Req2 *Minimizing Overfitting and Discovering General Behavioral Rules*: A model is said to be a good machine learning model if it generalizes any new input data from the problem domain. A pattern or rule is called general if it covers a relatively large subset of a relevant dataset. According to Geng et al. [32], generality measures the comprehensiveness of a pattern, that is, the fraction of all the relevant records in the dataset that matches the pattern. If a pattern characterizes more information in the relevant dataset, it tends to be more interesting [1, 32]. The main benefit of *generalization* in mining user behavioral rules is that it takes into account a minimal number of relevant contexts of users while producing rules. For example, typically a mobile phone user rejects most of the incoming calls when she is in a meeting, particularly calls from several relationships such as her friends, colleagues, or unknown. Such behavior of that individual is considered as "general behavior" during the meeting, which can be used to discover a general behavioral rule with high confidence. Unlike a traditional decision tree [2], where rules are extracted for each relationship available in the dataset, such general rule with high confidence plays an important role in effective behavior modeling for unseen test cases. This generalization not only simplifies the tree-based model but also minimizes the *over-fitting problem* of the traditional decision tree [2]. In our rule mining approach, we take into account such generalization while producing nodes according to an individual's behavior for a particular confidence level, to minimize the over-fitting problem and to discover the general behavioral rules of an individual user. A classification algorithm is said to over-fit, if it has lacking generalization, i.e., if it generates a decision tree utilizing the training dataset, which depends too much on irrelevant features of that data instances; It performs well on the training dataset but getting relatively poor performance to make predictions on unseen test cases. As the main purpose of a rule-based machine learning model is to make predictions on future data, which the model has never seen, the behavioral rule mining approach should have the ability to produce the *general* behavioral rules according to individual's preference.

Req3   *Discovering Specific Exception Rules and Non-redundancy*: Specific exception is another type of rule which represents different behavior of the corresponding general rule. Let's consider the above example, though the user rejects most of the incoming calls in a meeting, the user 'accepts' a few numbers of calls, particularly calls from her boss or mother. As such, the above general behavioral rules do not apply to her boss or mother. Thus, in addition to the above general behavioral rules $(meeting \Rightarrow reject)$, such type of specific exceptions $(meeting, boss \Rightarrow accept)$ or $(meeting, mother \Rightarrow accept)$ are needed to be discovered to model individual's behavior more effectively. As we discover only the specific exceptions of the relevant general behavioral rules, such discovery not only makes the rules useful but also avoids the redundant production in rules. Redundancy affects the rule quality and usefulness in practicality and makes the size of the discovered rule-set unnecessarily larger. This larger set is difficult to interpret and to identify the interesting rules for decision-makers and consequently makes the decision-making process ineffective and more complex. In our rule mining approach, we take into account the issue of redundancy while generating rules. Redundant production also takes a huge amount of time [9] and makes the overall process inefficient for real-world purposes. Thus, the behavioral rule mining approach should have the ability to produce the *specific exceptions* of the general behavioral rules, which will be *non-redundant* to provide real-time, proactive, and personalized services more effectively.

Req4   *Conflict Resolution in Multiple Rules*: Conflict in rules occurred when multiple rules provided different behavior for a particular context, associated in both rules. Let's consider the above phone call behavior example. For the same context 'meeting', the behavior of the general rule is 'reject', and the behavior of the specific exception rule is 'accept'. As the contexts in general rules (e.g., meeting) are also available in the corresponding specific exceptions (e.g., meeting, boss) or (e.g., meeting, mother), a rule order is needed to set the priority of the rules during the context 'meeting' in evaluation. However, the traditional confidence and support-based rule ordering [1] is not sufficient to predict the behavior for a particular test case. In addition to the support and confidence, the *number of relevant contexts* in the rule's antecedent also need to be considered for the ordering of rules to make conflict resolution. In our approach, we do ordering the discovered rules after extracting them from the contextual datasets. Thus, the behavioral rule discovery approach should have the ability to order the rules for handling the conflicts in rules.

In this work, we address the above four aspects for mining the *useful behavioral rules* of individual mobile phone users based on multi-dimensional contexts, to make the behavior modeling approach more effective, i.e., improving the accuracy for predicting individual mobile phone user's behavior.

6.4 Rule Mining Methodology

In this section, we present a rule-based machine learning approach for identifying behavioral association rules of individual users using their phone log dataset to provide context-aware intelligent services. Our method comprises multiple processing phases, which are detailed below.

6.4.1 Identifying the Precedence of Contexts

We choose the most appropriate context based on its precedence because different contexts, such as temporal, spatial, or social environment, may have varied implications on behavioral rules. For example, an incoming phone call from a significant person (e.g., mother) is usually accepted by an individual, although she may decline the call in other situations, e.g., if *social relationship* → *mother* then the *behavior* → *accept*. In this case, the relevance of a "social relationship" in making a call handling decision outweighs other factors such as time, location, and so on. The role of contexts in making decisions may differ for another individual, depending on her behavioral patterns in various situations.

To establish the precedence of contexts, we calculate entropy [33], which is a measure of impurity, and information gain [33], which splits the training samples into specific behavioral classes for a certain context. The highest precedence context is the one that has the most information. We first define entropy before we can quantify information gain properly. Entropy is a measure of impurity or disorder. The impurity of an arbitrary collection of samples is defined by the entropy. When the uncertainty is at its highest, it reaches its peak, and vice versa. Entropy and information gain are defined as follows [33]

$$H(S) = - \sum_{x \in X} p(x) log_2 p(x) \tag{6.1}$$

$$IG(A, S) = H(S) - \sum_{t \in T} p(t) H(t) \tag{6.2}$$

Where, $H(S)$ represents the entropy of set S, T represents the subsets created from splitting set S by attribute A such that $S = \cup_{t \in T} t$, $p(t)$ is the proportion of the number of elements in t to the number of elements in set S, and $H(t)$ represents the entropy of the subset t. In our tree-based method, the context containing the most information, such as temporal or spatial, or social, is regarded the highest precedence context in a given scenario. Consider the following sample dataset, which includes three different situations and the accompanying call response behavior of a mobile phone user X. According to her behavioral patterns, the contexts may be ranked as follows:

$Rank1 : Social\ Activity/Situation(S) \in \{meeting, lecture, seminar\}$
$Rank2 : Social\ Relationship(R) \in \{boss, mother, colleague, friend,$
$unknown\}$
$Rank3 : Temporal(T) \in \{time\text{-}of\text{-}the\text{-}week\}$

where,

$User\ phone\ call\ response\ behavior(BH) \in \{accept, reject\}$

6.4.2 Designing Association Generation Tree

In our tree based approach, we design an association generation tree (AGT), which is a common tree structure with a root node, several branches, and several interiors and/or leaf nodes. The tree is built from the root node, and each branch represents a context-aware test on a specific context value, such as in a meeting. Each node, whether interior or leaf, represents the corresponding outcome, which includes the behavioral activity class and the test's confidence value.

We use a top-down strategy to develop the AGT, which is one of the most common information processing methodologies. According to the precedence stated in the previous section, the tree is partitioned into behavioral activity classes distinguished by the values of the most relevant contexts. Once the tree's root node has been identified, the child nodes and their related branches can be discovered. The branches with the associated contexts and corresponding behavioral activity class indicated by the dominant behavior, i.e., indicating the most occurrences for a certain context, with a corresponding confidence value, are then recursively added to the tree. We consider the generalized pattern based on one's behavioral patterns for a given confidence level, say $t = 80\%$, that is preferred by individuals and may vary from user to user when generating nodes. After that, we compare the node to its parent nodes to see whether it contains redundant information. If it is detected, it is labeled REDUNDANT' ($NodeType \rightarrow$ 'REDUNDANT'). If both the child node and its parent node include the same behavior class and satisfy the individual's preferred confidence threshold, the node is considered redundant. Figure 6.1 depicts an example of AGT using several types of generated nodes

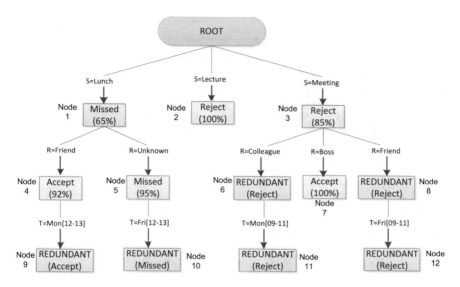

Fig. 6.1 An example of association generation tree (AGT) identifying 'REDUNDANT' nodes based on multi-dimensional contexts

and a confidence preference of 80%. Algorithm 3 describes the whole process of constructing AGT.

6.4.3   Extracting Non-redundant Behavioral Association Rules

Rules are extracted by traversing the tree while taking into account the decision nodes once the association generation tree (AGT) has been completed. To do so, we first choose the decision-making nodes from among all the tree's created nodes. If a node meets an individual's preferred confidence threshold ($NodeConf \geq t$) and is not labeled as 'REDUNDANT' ($NodeType \neq$ 'REDUNDANT') in the previous step, it is considered a decision node. As shown in Fig. 6.1, the decision nodes contain the behavioral activity classes and related confidence values satisfied by the user desired threshold ($t = 80\%$). As a result, Table 6.1 provides a summary of the AGT's produced nodes, as illustrated in Fig. 6.1.

We traverse from the root node to each decision-making node to construct non-redundant behavioral association rules that meet our objectives. The following are some instances of behavioral association rules generated by the tree.

R_1 : $Lecture \Rightarrow Reject$ (conf = 100%, using Node 2)
R_2 : $Meeting \Rightarrow Reject$ (conf = 85%, using Node 3)
R_3 : $Lunch, Friend \Rightarrow Accept$ (conf = 92%, using Node 4)
R_4 : $Lunch, Unknown \Rightarrow Missed$ (conf = 95%, using Node 5)
R_5 : $Meeting, Boss \Rightarrow Accept$ (conf = 100%, using Node 7)

Algorithm 3: Association generation tree

Data: Dataset: $DS = X_1, X_2, ..., X_n$ // each instance X_i contains a number of nominal
     context-values and corresponding behavior class BH, confidence threshold $= t$

Result: An association generation tree

1 <u>Procedure AGT</u> $(DS, context\_list, BHs)$;
2 $N \leftarrow createNode()$ //create a root node for the tree
3 **if** *all instances in DS belong to the same behavior class BH* **then**
4 | return N as a leaf node labeled BH with 100% confidence.
5 **end**
6 **if** *context_list is empty* **then**
7 | return N as a leaf node labeled with the dominant behavior class and corresponding
     confidence value.
8 **end**
9 identify the highest precedence context C_{split} for splitting and assign C_{split} to the node N.
10 **foreach** *context value val* $\in C_{split}$ **do**
11 | create subset DS_{sub} of DS containing val.
12 | **if** $DS_{sub} \neq \phi$ **then**
13 | | identify the dominant behavior and calculate the confidence value.
14 | | create a child node with the identified dominant behavior.
15 | | //check with its parent node
16 | | **if** *both nodes satisfy the confidence threshold* **then**
17 | | | **if** *both nodes represent same behavior class* **then**
18 | | | | //label the child node as 'REDUNDANT' node
19 | | | | $NodeType \rightarrow 'REDUNDANT'$
20 | | | **end**
21 | | **end**
22 | | add a subtree with new node and associated context values.
23 | | //recursively do this with remaining contexts
24 | | $AGT(DS_{sub}, \{context\_list - C_{split}\}, BHs))$
25 | **end**
26 **end**
27 return N

Table 6.1 An overview of the generated nodes in AGT shown in Fig. 6.1

Node no	Node parameters	Decision making summary
Node 01	$NodeConf = 65\% < t$	No (does not satisfy the confidence preference)
Node 02	$NodeConf = 100\% \geq t$	Yes (satisfy the condition of producing rule)
Node 03	$NodeConf = 85\% \geq t$	Yes (satisfy the condition of producing rule)
Node 04	$NodeConf = 92\% \geq t$	Yes (satisfy the condition of producing rule)
Node 05	$NodeConf = 95\% \geq t$	Yes (satisfy the condition of producing rule)
Node 06	$NodeType = $ 'REDUNDANT'	No (does not satisfy the node type)
Node 07	$NodeConf = 100\% \geq t$	Yes (satisfy the condition of producing rule)
Node 08	$NodeType = $ 'REDUNDANT'	No (does not satisfy the node type)
Node 09	$NodeType = $ 'REDUNDANT'	No (does not satisfy the node type)
Node 10	$NodeType = $ 'REDUNDANT'	No (does not satisfy the node type)
Node 11	$NodeType = $ 'REDUNDANT'	No (does not satisfy the node type)
Node 12	$NodeType = $ 'REDUNDANT'	No (does not satisfy the node type)

Rule R_1 states that the user always rejects incoming calls (100%), when the user is in a lecture, which is produced from node 2 in the tree. When the user is at a meeting (node 3 in the tree), rule R_2 indicates that the user rejects the majority of incoming calls (85%). The rule R_5 represents an exception to the general rule R_2, whereas this rule indicates the user's general behavior at a meeting. The user always (100%) takes her boss's calls while in a meeting, according to the specific exceptional rule R_5. If the related general behavior rule is not applicable in the appropriate contexts, specific exception rules play a vital role in modeling individual behavioral activities. Similarly, the rules R_3 and R_4 indicate user behavior at lunch, which can differ from one user to another in the real world.

6.5 Experimental Analysis

6.5.1 Effect on the Number of Produced Rules

In this experiment, we compare the discovered association rules utilizing both our rule discovery strategy and the traditional association rule mining-based techniques (BM). Apriori [1] is the most popular and widely utilized association rule mining approach in numerous application fields in the area of mining frequent patterns and association rules. Another prominent approach of mining frequent itemsets and rules, similar to the Apriori approach, is frequent pattern growth [33]. This method focuses on developing rules rather than candidates, which lowers database searching when generating rules. However, in terms of found rules, it yields a similar result.

Figures 6.2, 6.3, 6.4, and 6.5 show the relative comparison of produced number of rules for different users utilizing their datasets respectively. For this, we examined multiple confidence criteria, ranging from 100% (highest) to 60% (minimum)

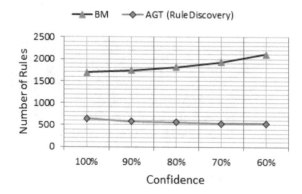

Fig. 6.2 Rule comparison for user U01

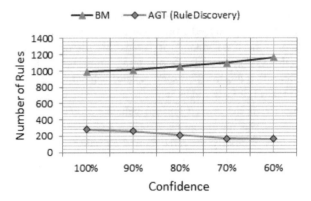

Fig. 6.3 Rule comparison for user U02

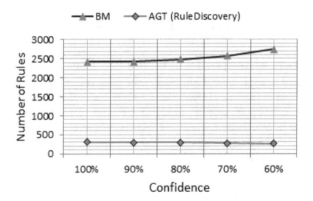

Fig. 6.4 Rule comparison for user U03

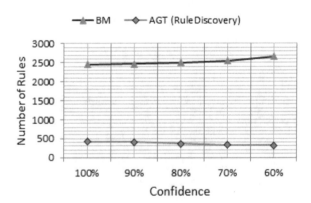

Fig. 6.5 Rule comparison for user U04

(lowest). We are not interested in using a confidence preference of less than 60% in our experimental study because confidence is directly associated with the accuracy of rules.

If we observe Figs. 6.2, 6.3, 6.4, and 6.5, we can see that as the confidence level is reduced, the number of identified association rules employing base approaches increases. The reason for this is that while producing rules, it simply takes into consideration all possible combinations of contexts. As a result, a lower confidence value satisfies more associations, resulting in a larger output. In our technique, however, for lower confidence thresholds, a greater proportion of child nodes subsume in their parent node, resulting in a reduction in the number of generated rules.

In terms of non-redundant rule discovery, if we further observe Figs. 6.2, 6.3, 6.4, and 6.5, we can see that our rule discovery methodology dramatically reduces the overall number of discovered rules, when compared to basic association rule mining approaches at different confidence criteria. The fundamental reason for this is that traditional association rule mining ignores redundancy analysis in associations when developing rules, resulting in an overly large rule set for a given confidence preference. On the other hand, we discover and eliminate redundancy while developing rules, extracting only non-redundant behavioral association rules.

As a result, our method minimizes the total number of identified rules while producing a compact collection of behavioral association rules. Such non-redundant behavioral association rule generation improves the effectiveness of our methodology and can be utilized to create a rule-based context-aware system that can quickly select appropriate rules from the identified rules and act accordingly. As a result of our experiments, we can conclude that our rule discovery methodology is capable of removing redundancy caused by existing base approaches while developing rules and providing a concise set of non-redundant behavioral association rules.

6.5.2 Effect of Confidence Preference on the Predicted Accuracy

We demonstrate the impact of confidence on precision and recall in this experiment. To do so, we first provide the detailed findings by adjusting the conference threshold for different individuals from 100% (highest) to 60% (lowest). Since confidence is defined as the accuracy of rules, we are not interested in using a value of less than 60% as a confidence preference. The relationship between precision and recall for different confidence criteria for two separate persons using their smartphone datasets is shown in Figs. 6.6 and 6.7.

Higher precision usually results to lower recall and vice versa. To make a reliable prediction of individuals, we apply a confidence preference in our methodology. In a context-aware test case, this allows users to set their preferences while making a decision. If we observe Figs. 6.6 and 6.7, we can see that as the confidence level is

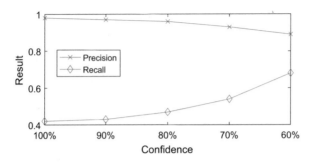

Fig. 6.6 Utilizing dataset of user U01

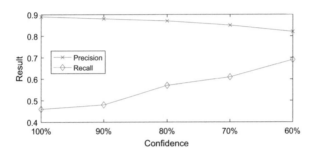

Fig. 6.7 Utilizing dataset of user U02

reduced, recall increases and vice versa. Precision, on the other hand, rises as the confidence threshold rises, and vice versa. The number of inaccurate predictions of context-aware test cases decreases as the confidence level rises, and the precision rises as a result. Overall, a greater confidence threshold leads to higher precision but poorer recall, while a lower confidence threshold leads to lower precision but higher recall, shown in Figs. 6.6 and 6.7 utilizing their datasets. A trade-off preference is more important in a user-centric application since the higher value of precision and recall shows the model efficacy. By default, we utilize an 80% confidence preference to find behavioral association rules for the users, assuming that all users have the same preference. We also give users the option of configuring the precision-recall trade-off based on their personal preferences. As a result, we may deduce that user-to-user confidence preferences have an impact on the anticipated accuracy of the resulting rule-based context-aware model.

6.5.3 Effectiveness Comparison

We demonstrate the usefulness of our rule-based model based on individual behavioral rules in terms of prediction accuracy in this experiment. As we begin by

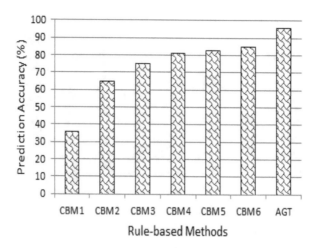

Fig. 6.8 Utilizing dataset of user U01

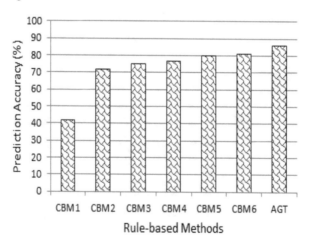

Fig. 6.9 Utilizing dataset of user U02

selecting numerous rule-based classification approaches from the machine learning and data science domains, such as ZeroR, OneR, RIDOR, DTNB, RIPPER, and DT. When compared to our rule-based model, these base classification approaches are represented as CBM1, CBM2, CBM3, CBM4, CBM5, and CBM6 accordingly in this experiment. As we discover rules based on the association generation tree, our rule discovery methodology is abbreviated as AGT. For a fair comparison, we utilize the same smartphone datasets in both training and testing sets. Figures 6.8 and 6.9 show the prediction accuracy for various rule-based classification methods of two different individual users in the area of our analysis.

Our technique significantly outperforms earlier rule-based classification systems for predicting individuals behavioral actions, as shown in Figs. 6.8 and 6.9. Aside

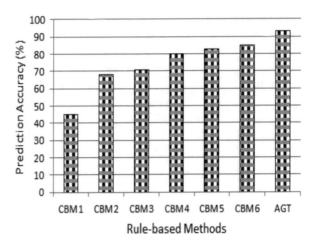

Fig. 6.10 Average accuracy comparison

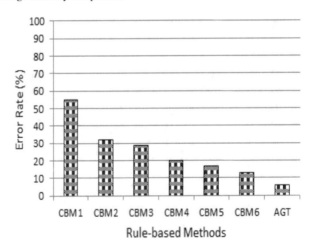

Fig. 6.11 Average error rate comparison

from individual comparisons, we also present a relative comparison of average prediction accuracy (percentage) and average error rate (percentage) for a collection of datasets, shown in Figs. 6.10 and 6.11 respectively. We do this by calculating the average prediction accuracy (percentage) and average error rate (percentage) using all of the datasets in this experiment. The average findings also reveal that our rule-based system consistently outperforms earlier methods in terms of higher accuracy and lower error rate. The reason is that this approach can effectively capture an individual's behavioral patterns in multi-dimensional contexts, which enhances the prediction outcomes. As a result, we may conclude that the presented rule discovery methodology is more effective in context-aware test cases than existing base models.

6.6 Conclusion

In this chapter, we have presented a rule-based machine learning method for detecting redundancy in associations and extracting a set of behavioral association rules based on context precedence. We created an association generation tree based on relevant multi-dimensional contexts such as temporal, spatial, and social factors to do this. The tree created in our method is used to extract non-redundant behavioral association rules. When compared to standard rule-based methods, our studies using people's contextual smartphone datasets reveal that this rule discovery methodology is more effective in terms of non-redundant rule generation and context-aware decision making. In this study, we also evaluated rule-based association and classification algorithms in terms of their capability to generate rules from a given dataset. The presented rule discovery approach will aid application developers in creating context-aware intelligent applications that will intelligently support smartphone users in their daily tasks. In addition to mobile applications, the concept of our rule mining method can also be applicable in different application domains, e.g., in the context of cybersecurity, this method can help to build machine learning rule-based intelligent security systems considering the surrounding contextual information, or the security features.

References

1. Agrawal, R., & Srikant, R. (1994, September). Fast algorithms for mining association rules. In *Proc. 20th Int. Conf. Very Large Data Bases, VLDB* (Vol. 1215, pp. 487–499).
2. Quinlan, J. R. (2014). *C4. 5: Programs for machine learning*. Elsevier.
3. Hong, J., Suh, E. H., Kim, J., & Kim, S. (2009). Context-aware system for proactive personalized service based on context history. *Expert Systems with Applications, 36*(4), 7448–7457.
4. Lee, W. P. (2007). Deploying personalized mobile services in an agent-based environment. *Expert Systems with Applications, 32*(4), 1194–1207.
5. Zulkernain, S., Madiraju, P., Ahamed, S. I., & Stamm, K. (2010). A mobile intelligent interruption management system. *Journal of Universal Computer Science, 16*(15), 2060–2080.
6. Mehrotra, A., Hendley, R., & Musolesi, M. (2016, September). PrefMiner: Mining user's preferences for intelligent mobile notification management. In *Proceedings of the 2016 ACM International Joint Conference on Pervasive and Ubiquitous Computing* (pp. 1223–1234).
7. Ordonez, C. (2006, November). Comparing association rules and decision trees for disease prediction. In *Proceedings of the International Workshop on Healthcare Information and Knowledge Management* (pp. 17–24).
8. Freitas, A. A. (2000). Understanding the crucial differences between classification and discovery of association rules: A position paper. *ACM SIGKDD Explorations Newsletter, 2*(1), 65–69.
9. Srinivasan, V., Moghaddam, S., Mukherji, A., Rachuri, K. K., Xu, C., & Tapia, E. M. (2014, September). Mobileminer: Mining your frequent patterns on your phone. In *Proceedings of the 2014 ACM International Joint Conference on Pervasive and Ubiquitous Computing* (pp. 389–400).

10. Zhu, H., Chen, E., Xiong, H., Yu, K., Cao, H., & Tian, J. (2014). Mining mobile user preferences for personalized context-aware recommendation. *ACM Transactions on Intelligent Systems and Technology, 5*(4), 1–27.
11. Fournier-Viger, P., Wu, C. W., & Tseng, V. S. (2012, May). Mining top-k association rules. In *Canadian Conference on Artificial Intelligence* (pp. 61–73). Berlin, Heidelberg: Springer.
12. Bouker, S., Saidi, R., Yahia, S. B., & Nguifo, E. M. (2012, November). Ranking and selecting association rules based on dominance relationship. In *2012 IEEE 24th international conference on tools with artificial intelligence* (Vol. 1, pp. 658–665). IEEE.
13. Halvey, M., Keane, M. T., & Smyth, B. (2006, April). Time based patterns in mobile-internet surfing. In *Proceedings of the SIGCHI Conference on Human Factors in Computing Systems* (pp. 31–34).
14. Sarker, I. H., Colman, A., Kabir, M. A., & Han, J. (2018). Individualized time-series segmentation for mining mobile phone user behavior. *The Computer Journal, 61*(3), 349–368.
15. Park, M. H., Hong, J. H., & Cho, S. B. (2007, July). Location-based recommendation system using Bayesian user's preference model in mobile devices. In *International conference on ubiquitous intelligence and computing* (pp. 1130–1139). Berlin, Heidelberg: Springer.
16. Zhang, L., Liu, J., Jiang, H., & Guan, Y. (2013). SensTrack: Energy-efficient location tracking with smartphone sensors. *IEEE Sensors Journal, 13*(10), 3775–3784.
17. Sun, X., & May, A. (2009). The role of spatial contextual factors in mobile personalization at large sports events. *Personal and Ubiquitous Computing, 13*(4), 293–302.
18. Eagle, N., & Pentland, A. S. (2006). Reality mining: sensing complex social systems. *Personal and Ubiquitous Computing, 10*(4), 255–268.
19. Farrahi, K., & Gatica-Perez, D. (2008, October). What did you do today? Discovering daily routines from large-scale mobile data. In *Proceedings of the 16th ACM International Conference on Multimedia* (pp. 849–852).
20. Sarker, H., Sharmin, M., Ali, A. A., Rahman, M. M., Bari, R., Hossain, S. M., & Kumar, S. (2014, September). Assessing the availability of users to engage in just-in-time intervention in the natural environment. In *Proceedings of the 2014 ACM International Joint Conference on Pervasive and Ubiquitous Computing* (pp. 909–920).
21. Sarker, I. H. (2019). Context-aware rule learning from smartphone data: Survey, challenges and future directions. *Journal of Big Data, 6*(1), 1–25.
22. Sarker, I. H., Colman, A., Han, J., Kayes, A. S. M., & Watters, P. (2020). CalBehav: A machine learning-based personalized calendar behavioral model using time-series smartphone data. *The Computer Journal, 63*(7), 1109–1123.
23. Rosenthal, S., Dey, A. K., & Veloso, M. (2011, June). Using decision-theoretic experience sampling to build personalized mobile phone interruption models. In *International conference on pervasive computing* (pp. 170–187). Berlin, Heidelberg: Springer.
24. Dekel, A., Nacht, D., & Kirkpatrick, S. (2009, September). Minimizing mobile phone disruption via smart profile management. In *Proceedings of the 11th International Conference on Human-Computer Interaction with Mobile Devices and Services* (pp. 1–5).
25. Sykes, E. R. (2014). A cloud-based interaction management system architecture for mobile devices. *Procedia Computer Science, 34*, 625–632.
26. Lee, D., Park, S. E., Kahng, M., Lee, S., & Lee, S. G. (2010). Exploiting contextual information from event logs for personalized recommendation. In *Computer and Information Science 2010* (pp. 121–139). Berlin, Heidelberg: Springer.
27. Bolger, N., Davis, A., & Rafaeli, E. (2003). Diary methods: Capturing life as it is lived. *Annual Review of Psychology, 54*(1), 579–616.
28. Thayer, A., Bietz, M. J., Derthick, K., & Lee, C. P. (2012, February). I love you, let's share calendars: Calendar sharing as relationship work. In *Proceedings of the ACM 2012 conference on Computer Supported Cooperative Work* (pp. 749–758).
29. Tungare, M., Perez-Quinones, M., & Sams, A. (2008). An exploratory study of calendar use. arXiv preprint arXiv:0809.3447.

30. Sarker, I. H., & Kayes, A. S. M. (2020). ABC-RuleMiner: User behavioral rule-based machine learning method for context-aware intelligent services. *Journal of Network and Computer Applications, 168,* 102762.
31. Khalil, A., & Connelly, K. (2006, November). Context-aware telephony: Privacy preferences and sharing patterns. In *Proceedings of the 2006 20th anniversary conference on computer supported cooperative work* (pp. 469–478).
32. Geng, L., & Hamilton, H. J. (2006). Interestingness measures for data mining: A survey. *ACM Computing Surveys, 38*(3), 9–es.
33. Witten, I. H., & Frank, E. (2002). Data mining: Practical machine learning tools and techniques with Java implementations. *ACM SIGMOD Record, 31*(1), 76–77.

Chapter 7
Recency-Based Updating and Dynamic Management of Contextual Rules

7.1 Introduction

Mobile phone log data is not static as it is progressively added to day-by-day according to individual's diverse behaviors with mobile phones. Since an individual's behavior changes over time, the most recent pattern is likely to be more significant than older ones for predicting individual mobile phone user behavior [1].

Currently, researchers use static periods of recent log data to produce rules that express users' present behavior. For example, Lee et al. [2] have studied the mobile phone users' calling patterns and used the last 3 months' call logs. Phithakkitnukoon et al. [3] have presented a model for predicting incoming and outgoing calls and assumed the latest 60 days call logs data to model future call activities. The problem of utilizing a static period of log data to produce rules is that those rules may not reflect the present behavior of a user, as an individual's behavior changes over time. Let's consider the last 3 months' call log data of a mobile phone user. Assume that as per log data the user has a call 'reject' behavioral pattern [10:00 AM–12:00 PM] as she used to have a regular meeting at that time. Recently, she has no meeting at that time period on Monday and she typically 'accepts' incoming phone calls. So for this example, the past 'reject' behavioral pattern, even with high evidence (support value) according to log data, is not meaningful to predict her future behavior. Therefore, we need to find out when the *behavior changes* of individual users' so that more currently relevant rules can be formulated.

To dynamically identify such period, if we assume only a short period (e.g., last week's data) as indicative of recent behavior, there may not be enough data instances in that period to infer a valid rule for predicting future phone call behavior. Creating rules based on observations with so little "support" is unlikely to be effective [4]. On the other hand, if we take into account a comparatively longer period (e.g., last 6 months data) as indicative of recent behavior, we could get greater support but it might result in a greater behavioral variation thus decrease the confidence of some expected rules. As a consequence, we may miss these rules because of low

© The Author(s), under exclusive license to Springer Nature Switzerland AG 2021
I. H. Sarker et al., *Context-Aware Machine Learning and Mobile Data Analytics*,
https://doi.org/10.1007/978-3-030-88530-4_7

confidence. Therefore, an optimal period of log data that reflects the recent behavior of an individual needs to be identified. Individuals' behavioral rules then need to be updated to remove outdated rules (out-of-date rules) that conflict with their recent behavior and to include more recent behavioral rules that do not exist in the existing rule-set, to get a complete set of significant rule-set for use in various real-world applications.

In this chapter, we present a recency-based updating approach that not only removes the outdated rules (rules that do not represent the present behavior of an individual) from the existing rule-set but also outputs a complete set of updated rules according to individuals' recent behavioral patterns. In this approach, we first dynamically identify an optimal period for which a recent behavioral pattern has been dominant by analyzing the behavioral characteristics of individual mobile phone users utilizing their mobile phone data. Once we have determined the recent log data, we then identify the outdated rules from the existing rule-set (discovered from entire log data), by checking the behavior for a particular context in both recent log data and the entire log data of an individual. After that, this recency-based approach removes the outdated rules from the existing rule-set and outputs a complete set of updated rules by merging the existing rule-set and the newly discovered rule-set from the recent log data.

7.2   Requirements Analysis

In this section, we discuss and summarize the key requirements of the recency-based approach. These are:

Req1   *Identifying Changes in Individual's Behavioral Patterns and Determining Recent Log:* One of the most important requirements for extracting an individual's recency-based behavioral rules is identifying changes in behavioral patterns and determining related dynamic recent log data. By examining an individual's behavioral patterns in relevant contexts, an appropriate period of recent log data that reflects their recent behavior can be determined. Individuals' recency-based rules can be discovered using this dynamic optimal period of data. Thus, the method should be able to detect the changes in an individual's behavioral patterns utilizing their phone log data without any prior knowledge.

Req2   *Detecting and Removing Outdated Rules:* An outdated (out-of-date) behavioral rule is a valid rule in terms of rule's constraints (e.g., support and confidence) but does not represent the recent behavior of an individual user. The definition of an outdated rule of individual mobile phone users is formally stated as—Let, a rule $R_1 : A_1 \Rightarrow C_1$ that is discovered from the entire mobile phone dataset DS, where A_1 represents the contextual information and C_1 is the mobile phone usage behavior. The rule R_1 is considered as an outdated rule $R_{outdated}$, if and only if C_1 is identified as conflict (different behavior) for that context A_1 utilizing recent phone log data DS_{recent}, i.e., $A_1 \Rightarrow C_2$ and $C_1 \neq C_2$, where C_1 and C_2 represent the past and recent behavior respectively for A_1. In general, this type of rule is produced based on past behavioral patterns of individuals utilizing the entire phone log data. As the most *recent pattern* is more significant than older ones, the outdated rules even with high support values increase the error rate for predicting an individual's future behavior. Therefore, the approach should have the ability to *detect and remove the outdated rules* from the rule-set extracted from entire phone log data.

Req3 *Discovering New Recent Behavioral Rules:* A new recent behavioral rule is a rule that is not produced when utilizing the entire phone log data DS but is produced when utilizing the recent period of log data DS_{recent}. The definition of a new recent behavioral rule of individual mobile phone users is formally stated as—Let, a rule $R : A \Rightarrow C$ that is produced utilizing recent log data DS_{recent}, where A represents the contextual information and C is the mobile phone usage behavior. The rule R is considered as a new recent behavioral rule R_{new}, if and only if, there is no such rule discovered from the entire log dataset DS. Although DS_{recent} is a subset of DS, such kinds of rules are not discovered utilizing the entire log data DS because of their low confidence value and not satisfying the user preferred confidence threshold (say, 80%). The reason is that an individual's behavior changes over time for a particular context and several variations in user's behavior or conflicts for that context decrease the confidence of the associated behavior. However, a strong behavioral pattern with high confidence may be found in the recent phone log DS_{recent}, which satisfies the user preferred confidence threshold. Such new rules make the behavior model more significant to predict an individual's future behavior. Therefore, the approach should have the ability to produce such *new recent behavioral rules* of individuals.

Req4 *Dynamic Management of Rules:* As the recency-based approach is responsible not only to identify the dynamic optimal period of recent log data but also for identifying and removing the outdated rules, and discovering new recent behavioral rules, dynamic management of rules is needed to get a complete set of updated rules without making any assumptions about when individual's behavior changed to a new pattern. Let, $RS_{initial}$ be a set of rules discovered from entire mobile phone data DS, and RS_{recent} be another set of rules discovered from recent log data DS_{recent}. A complete set of recency based updated rules $RS_{updated}$ will be the merging output of these two rule-sets, e.g., $RS_{updated} = merge(RS_{initial}, RS_{recent})$. This complete updated rule-set $RS_{updated}$ not only contains all the significant rules of an individual mobile phone user but also expresses recent behavioral patterns that will be applicable for modeling mobile phone usage behavior in real-world applications.

7.3 An Example of Recent Data

The concept of recent log data is formally stated as—Let, s_1 be the number of instances (records) in the entire mobile phone dataset DS, which is temporally ordered. A recent mobile phone dataset DS_{recent} is a subset of DS, which contains the most recent records of DS based on timestamps of size s_2, where $s_2 \leq s_1$.

Table 7.1 Sample recent log dataset (from record r_s to r_n) containing phone call activities with multi-dimensional contexts

Record No	Temporal	Location	Social activity (Situation)	Social relationship	User activity
r_1	Mon[10:00–12:00]	Office	Meeting	Friend	Reject
r_2	Mon[10:00–12:00]	Office	Meeting	Boss	Accept
r_3	Mon[10:00–12:00]	Office	Meeting	unknown	Reject
r_4	Mon[18:30–19:30]	Home	Dinner	Friend	Accept
r_5	Tue[18:30–19:30]	Home	Dinner	Friend	Accept
r_6	Tue[18:30–19:30]	Home	Dinner	Friend	Accept
r_7	Fri[10:15–11:30]	Office	Lecture	Friend	Reject
r_8	Fri[14:30–15:30]	Office	Lab	Friend	Accept
r_9	Mon[10:00–12:00]	Office	Meeting	Friend	Reject
r_{10}	Mon[18:30–19:30]	Home	Dinner	unknown	Missed
–	–	–	–	–	–
–	–	–	–	–	–
–	–	–	–	–	–
	Recent data				
r_s	–	–	–	–	–
–	–	–	–	–	–
–	–	–	–	–	–
r_{n-3}	Mon[10:00–12:00]	Office	No event	Friend	Accept
r_{n-2}	Wed[10:00–12:00]	Office	Meeting	Friend	Reject
r_{n-1}	Fri[10:15–11:30]	Office	Seminar	Colleague	Accept
r_n	Fri[14:30–15:30]	Office	Lab	Friend	Reject

Table 7.1 shows an example of recent log DS_{recent} starting from the record r_s to r_n, which is a subset of the entire log data DS starting from the record r_1 to r_n ordered temporally. It reports some pieces of information coming from a phone call log that records user phone call activities with corresponding context values. For each record, temporal, locational, and social information, and user corresponding phone call activity are stored.

Each record in the dataset (see Table 7.1) is ordered temporally, i.e., ordered according to the temporal information sequentially. For instance, a record r_1 represents an activity on Monday between 10:00 AM and 12:00 PM of a particular week. Record r_5 represents another activity of the next day Tuesday between 18:30 and 19:30 of that week. Similarly, a record r_9 represents another similar activity (meeting) of Monday between 10:00 AM and 12:00 PM of the next week. According to Table 7.1, the record r_n is the most recent activity of an individual user and the record r_s represents the starting record of the recent log, i.e., the behavioral patterns based on relevant contexts before r_s are considered as past behavior and the behavioral patterns after r_s up to r_n are considered as recent behavior of users. If there is no change in behavioral patterns from the record r_1 (beginning of entire log data) to r_n, then the behavioral patterns for the entire log are considered as recent patterns. We utilize such variable length of recent log data in our recency-based approach to discover the recent behavioral patterns more properly. It varies from user to user depending on how the user's behavior changes over time of the week in different contexts.

7.4 Identifying Optimal Period of Recent Log Data

Identifying an optimal period of recent behavioral data dynamically from the entire phone log is the key to our recency-based approach for individual mobile phone users. This section presents all the phases to identify such a period of log data.

7.4.1 Data Splitting

We split the entire log into week-by-week data in this initial phase since the time of the week is the most important factor influencing user behavior [5]. We chose weekly basis splitting since no one's behavior is likely to be the same every day of the week (Monday, Tuesday,..., Sunday). As a result, we expect weekly patterns of behavior to repeat (e.g., a user has the same days off work each week). Week $W1$ represents the initial week data in the mobile phone log of an individual mobile phone user, while week Wn represents the most recent week data in the mobile phone log of an individual mobile phone user as shown in Fig. 7.1.

7.4.2 Association Generation

Once the data splitting has been completed, we generate context-association for each set of week-wise data DS_{week} starting from the most recent week W_n. Context association is simply the combination of contexts, where—

(i) the association may contain single (user social activity, e.g., meeting) or multi-dimensional contexts (user social activity, e.g., meeting, user location, e.g., office).

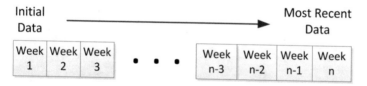

Fig. 7.1 An example of week-wise data splitting

(ii) contexts are added incrementally according to the precedence of contexts to create an association based on multi-dimensional contexts.
(iii) each context may occur at most once in an association.
(iv) the number of contexts in an association is less or equal to the total number of contexts in a given dataset DS_{week}.

We calculate information gain [6], which is a statistical property that generates entropy and measures how well a particular context-value separates the training datasets into specific behavior classes available in the dataset, to discover the precedence of contexts in a dataset. The highest precedence context is defined as the one with the largest information gain value. The process for generating context associations is set out in Algorithm 4. Input data includes week wise data: $DS_{week} = X_1, X_2, ..., X_n$, which contains a set of instances with categorical contexts and output data is the association list $assoc\_list$. The algorithm returns the generated association list $assoc\_list$. A combination {office, meeting} is an example of context association containing 2-contexts.

Algorithm 4: Context association generation

Data: Week-wise dataset: $DS_{week} = X_1, X_2, ..., X_n$ // each instance X_i contains a number of categorical contexts
Result: association list $assoc\_list$

1 $initial\_assoc \leftarrow \phi$; // initialization
2 Procedure CAG ($DS_{week}, context\_list$);
3 **if** $context\_list$ is empty **then**
4 | return $assoc\_list$;
5 **end**
6 //calculate entropy and information gain for each context and identify the precedence of contexts
7 $A_{best} \leftarrow$ the context that best classifies examples;
8 **foreach** context value $con \in A_{best}$ **do**
9 | $assoc \leftarrow create\_association(con, initial\_assoc)$;
10 | added $assoc$ to the assoc_list;
11 | // partition into subsets and grow the corresponding subtrees for each subset
12 | $DS_{sub} \leftarrow$ subset of DS_{week} that have value con;
13 | **if** DS_{sub} is empty **then**
14 | | $initial\_assoc \leftarrow \phi$;
15 | **else**
16 | | //remove A_{best} context from the context list
17 | | $context\_list \leftarrow context\_list - A_{best}$;
18 | | // recursively generated association with remaining attributes
19 | | $CAG(DS_{sub}, context\_list)$
20 | **end**
21 **end**

7.4.3   Score Calculation

We calculate the conflict score based on the conflict behavior for each association between two adjacent weeks after producing the context associations. We start by identifying the dominant behavior [4] because we don't always expect a user to behave exactly like another for a given association.

For example, if a user rejects 85% of incoming calls, accepts 10%, and misses 5% of incoming calls for a specific association of context (e.g., meeting, office), reject will be the dominant behavior for that association. For example, if a user accepts 65% of incoming calls and rejects 35% for a specific association of context (e.g., seminar, office, coworker), then acceptance will be the dominant behavior for that association. We start scanning from the most recent week W_n and continue to all previous weeks W_{n-1}, W_{n-2}, W_{n-3}, ..., W_1 one by one to identify the conflict behavior for each context association in the adjacent weeks.

We calculate the conflict score using Eq. 7.1 after determining whether or not there is a conflict for each context association established in the previous section. If $assoc_{total}$ represents the total number of associations generated in a week W_n and $conflict_{total}$ represents the total number of conflicts discovered when comparing the generated associations in a week W_n and the adjacent week $[W_{n-1}]$, then the percentage (percentage) of conflict score with respect to the most recent week W_n is defined as follows:

$$score(\%) = \frac{conflict_{total}}{assoc_{total}} \times 100 \tag{7.1}$$

The process for calculating this conflict score is set out in Algorithm 5. Input data includes adjacent weeks data: DS_{week1} for a week W_n and DS_{week2} for a week W_{n-1}, each of which contains a set of training instances X_1, X_2, ..., X_n, and output data is the conflict score in percentage. We first generate context associations for DS_{week1} and DS_{week2} using Algorithm 4. After that for each association, we check whether the dominant behavior is the same or not. If different dominant found then the number of conflict increases. After that, we calculate the percentage (%) of conflict behaviors. Finally, this algorithm returns the calculated score.

7.4.4   Data Aggregation

The final stage in determining an optimal period of recent log data is data aggregation. We do this by aggregating the week-wise data based on similar behavioral patterns identified by conflict score. As we do not expect the same contextual information in each week, we use the conflict score between two adjacent weeks rather than likelihood to assess behavioral similarity. For example, the user may attend a seminar during one week but not throughout all weeks. The conflict score, on the other hand, identifies behavioral differences between 2 weeks. If the conflict

---

Algorithm 5: Conflict score calculation

Data: adjacent weeks data: DS_{week1} for week W_n and DS_{week2} for week W_{n-1}.
Result: conflict score: *score*

1 *assoc_list1 $\leftarrow$ generate_association(DS_{week1})*;
2 (using algorithm 4)
3 *assoc_list2 $\leftarrow$ generate_association(DS_{week2})*;
4 (using algorithm 4)
5 //initializing score count
6 initialize *count* $\leftarrow 1$
7 **foreach** *association assoc in assoc_list1* **do**
8   **if** *same assoc found in assoc_list2* **then**
9   //identify the dominant behavior for assoc
10       $BH1 \leftarrow identify\_dominant(assoc, DS_{week1})$;
11       $BH2 \leftarrow identify\_dominant(assoc, DS_{week2})$;
12   //check the dominant behavior
13   **if** $BH1 \neq BH2$ **then**
14 | increment *count*
15   **end**
16   **end**
17 **end**
18 //calculate the percentage (%) of conflicts score
19 *score $\leftarrow$ calculatePercentage(count, assoc_list1)*
20 return *score*

---

score of two consecutive weeks is 0% (no conflict), the behavioral patterns in these 2 weeks are very similar [7]. We aggregate the conflict scores of two adjacent weeks from the most recent week W_n to the prior weeks $[W_{n-1}, W_{n-2}, ...,]$ until we achieve a substantial difference in the conflict scores of two adjacent weeks. Then, for recent similar behavioral patterns, we established a boundary line. A significant variation is encountered when it exceeds the average result of the variations by considering the overall behavior in the entire datasets. If S_{total} is the total conflict score and N_{weeks} is the overall number of weeks in a dataset, the average score is as follows:

$$average\ score = \frac{S_{total}}{N_{weeks}} \tag{7.2}$$

This aids in determining an optimal period of recent log data by identifying the dynamic threshold rather than assuming a static threshold. According to their behavioral consistency, such a threshold may differ from user to user. Thus, recent behavioral patterns are discovered for some users by averaging a large number of weeks and for others by aggregating a smaller number of weeks, based on how the user's behavior changes over time of the week in various contexts.

Figure 7.2 shows an example of recent log data by aggregating the most recent 4 weeks data (from Week W_{n-3} up to Week W_n), which reflect the recent behavioral patterns of an individual user. According to Fig. 7.2, week W_n is the most recent

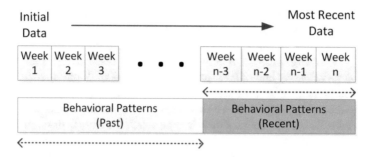

Fig. 7.2 An example of data aggregation for identifying recent log data

week and week W_{n-3} is the boundary of recent behavioral patterns, that is, the behavioral patterns based on related contexts before week W_{n-3} (from week W_1 up to week W_{n-3}), are considered as past behavior and the behavioral patterns after week W_{n-3} up to the most recent week W_n (from week W_{n-3} up to week W_n), are considered as recent behavior of that user. If there is no change in behavioral patterns from week W_1 (beginning of log data) up to week W_n, then the behavioral patterns in the entire log data are considered as recent patterns. Our approach dynamically creates an optimal period of recent log data from an individual's cell phone data, rather than arbitrarily determining the number of periods in advance. As a result, the number of weeks and time bounds for recent log data will vary for each user, depending on how the user's behavior varies through time and context. We utilize such variable length of recent log data for producing individual recency-based rules.

7.5   Machine Learning Based Behavioral Rule Generation and Management

We create rules based on the recent log data $DSrecent$ once it has been determined. On recent log data, we use our previous rule-based machine learning method [8], association generation tree, to generate rules. The reason for using tree-based learning is that the nodes closer to the root are more general, which can be utilized to mine broader behavioral patterns. To build behavioral rules, this method first creates a tree based on context precedence, with each node representing a behavior class and its accompanying confidence value. Rules are extracted by traversing the tree from the root node to each decision node, indicated by the node's value, once the tree has been designed. To simulate individual mobile phone user behavior, this methodology generates a set of human-understandable behavioral rules based on multi-dimensional contexts. When more context dimensions are taken into account, the created rules not only capture an individual's generic behavior at a certain level of confidence with a small number of contexts but also express specific exceptions to the general rules. When a user is at a meeting, for example, she normally rejects the majority of incoming calls (83%); but, she always accepts (100%) if the incoming call is from her boss.Thus the produced general and specific exception rule are represented as $R_{general}$: $meeting \Rightarrow reject$ (conf = 83%) and $R_{exception}$: $meeting, boss \Rightarrow accept$ (conf = 100%) respectively. Such produced rules are *non-redundant* and *reliable* according to individual's preferred confidence.

In our approach, once we have produced rules utilizing a dynamic length of recent log data DS_{recent}, we merge this rule-set with initial rule-set $RS_{initial}$ that is produced utilizing the entire phone log data DS. To extract the initial rule-set $RS_{initial}$, we also use the same rule discovery approach [8] discussed above, in order to output a *complete set of updated rules* $RS_{updated}$ for each individual user. While merging, we identify and remove the *outdated rules* from the initial rule-set $RS_{initial}$, as these rules do not represent the recent behavior of an individual. We also remove rules from RS_{recent} that exist in the initial rule set $RS_{initial}$. Thus, we output a complete set of recency-based updated rules by taking into account the behavioral patterns in both the rule-sets $RS_{initial}$ and RS_{recent} using a rule merging operation, e.g., $RS_{updated} = merge(RS_{initial}, RS_{recent})$.

7.6   Effectiveness Comparison and Analysis

We compare the effectiveness of our recency-based modeling for providing context-aware mobile services to the base model in terms of prediction coverage and error rate [1]. Figures 7.3 and 7.4 show the above-mentioned error rate (%) and coverage

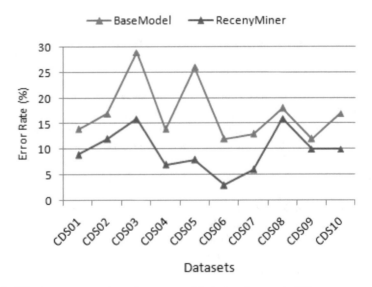

Fig. 7.3 Effectiveness comparison of our approach in terms of error rate (%)

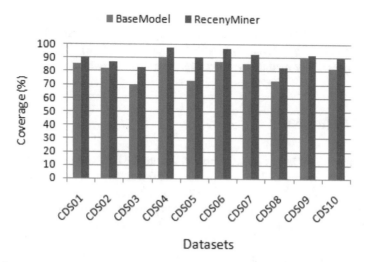

Fig. 7.4 Effectiveness comparison of our approach in terms of prediction coverage (%)

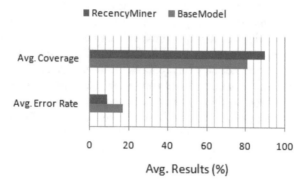

Fig. 7.5 Effectiveness comparison of our approach in terms of average prediction coverage (%) and error rate (%) of all datasets

(%) for different users utilizing their own datasets. For a particular confidence preference of 80%, the results are provided for both procedures.

Our recency-based approach consistently outperforms the base model for predicting individual mobile phone usage behavior, as shown in Figs. 7.3 and 7.4. The fundamental reason for this is that rules generated by the base model do not represent the rule's freshness according to the recent behavior of individuals. Our recency-based solution, on the other hand, tackles this problem by generating rules based on an individual's recent behavioral patterns, making the technique more useful by maximizing prediction coverage as well as minimizing error rate.

In addition to individual comparisons, we show a relative comparison of average prediction coverage and error rate in predictions for a collection of datasets when compared to the base model. Figure 7.5 shows the average results (average prediction coverage and average error rate considering all datasets). For a collection of datasets, the average findings suggest that our recency-based strategy outperforms the base model. The reason is that when we create behavioral rules for specific users, we take into account their recent behavioral patterns, which improves the effectiveness of our method by better capturing their behavioral patterns.

7.7 Conclusion

In this chapter, we have discussed how to update the behavioral rules of individual mobile phone users based on current behavioral patterns as well as how to manage them dynamically. We took into account four aspects for this, including detecting changes in individual behavior and determining an optimal period of recent log data, identifying and removing outdated rules that do not represent an individual's recent behavior, discovering new recent behavioral rules that do not exist in the existing rule-set, and dynamic management of these rules to get the best results. A method for identifying an optimal period of recent log data from the entire log data, which is the key of the recency-based behavior modeling has also been presented. The primary theme of this methodology is that we dynamically identify such periods by comparing the behavioral similarities of individual mobile phone users across time, which can vary from user to user in the real world. The resultant updated rule set produced by this model not only incorporates all the significant rules of the users but also reflects their recent behavioral patterns in rules that can play a significant role in a variety of real-world applications. In addition to mobile applications, the concept of the recency-based model can also be applicable in different application domains, e.g., in the context of cybersecurity this can help to effectively analyze the effect of cyber-attacks based on their recent behavioral patterns in cyberspace to build better secured systems.

References

1. Sarker, I. H., Colman, A., & Han, J. (2019). Recencyminer: mining recency-based personalized behavior from contextual smartphone data. *Journal of Big Data, 6*(1), 1–21.
2. Lee, S., Seo, J., & Lee, G. (2010, April). An adaptive speed-call list algorithm and its evaluation with ESM. In *Proceedings of the SIGCHI Conference on Human Factors in Computing Systems* (pp. 2019–2022).
3. Phithakkitnukoon, S., Dantu, R., Claxton, R., & Eagle, N. (2011). Behavior-based adaptive call predictor. *ACM Transactions on Autonomous and Adaptive Systems (TAAS), 6*(3), 1–28.
4. Sarker, I. H., Colman, A., Kabir, M. A., & Han, J. (2018). Individualized time-series segmentation for mining mobile phone user behavior. *The Computer Journal, 61*(3), 349–368.
5. Halvey, M., Keane, M. T., & Smyth, B. (2006). Time based patterns in mobile-internet surfing. In *Proceedings of the SIGCHI Conference on Human Factors in Computing Systems* (pp. 31–34).
6. Quinlan, J. R. (2014). *C4. 5: programs for machine learning.* Elsevier.

7. Sarker, I. H., Kabir, M. A., Colman, A., & Han, J. (2017, November). Identifying recent behavioral data length in mobile phone log. In *Proceedings of the 14th EAI International Conference on Mobile and Ubiquitous Systems: Computing, Networking and Services* (pp. 545–546).
8. Sarker, I. H., & Kayes, A. S. M. (2020). ABC-RuleMiner: User behavioral rule-based machine learning method for context-aware intelligent services. *Journal of Network and Computer Applications, 168*, 102762.

Part III
Rule-Based Systems, Deep Learning and Challenges

This part of the book discusses machine learning rule-based expert system modeling (Chap. 8), deep learning modeling for context-aware systems (Chap. 9), and finally a conclusion recapping the contributions of this book, real-world applications, and major challenges and research issues for future investigation (Chap. 10).

Chapter 8
Context-Aware Rule-Based Expert System Modeling

8.1 Structure of a Context-Aware Mobile Expert System

A mobile expert system modeling is based on a set of context-aware rules to build various smart mobile applications. The simplest type of AI method that uses prescribed knowledge-based rules to solve an issue is usually a rule-based expert system [1]. Usually, the purpose of the expert system is to take information from a human expert and turn this into a number of hardcoded rules for the input data to be implemented. In this chapter, we focus on the generated rules based on machine learning techniques, rather than the hardcoded rules, as we take into account the dynamism in the context-aware rules and corresponding applications. In the following, we first discuss about the structure of a rule-based mobile expert system, and then we discuss how we can use context-aware rules extracted using machine learning techniques for the rule-based expert system.

An example of a knowledge-based system is a mobile expert system, which is divided broadly into two subsystems, such as the inference engine and the knowledge base, shown in Fig. 8.1. In addition, a user interface is another part of a complete expert system. In the following, we discuss the roles of each module, shown in Fig. 8.1.

- *Knowledge-base:* As it consists of knowledge of the target mobile domain as well as operational knowledge of the decision rules of apps, the knowledge base module is the basis of an expert system. The knowledge base is a collection of rules or other information structures derived typically from the human expert. As we are interested on data-driven automated rules, rather than the typical hardcoded rules, we take into account the rule-based machine learning methods to generate rules. Rules are typically structured as IF-THEN statements of the form:

  IF $<antecedent>$ THEN $<consequent>$

© The Author(s), under exclusive license to Springer Nature Switzerland AG 2021
I. H. Sarker et al., *Context-Aware Machine Learning and Mobile Data Analytics*,
https://doi.org/10.1007/978-3-030-88530-4_8

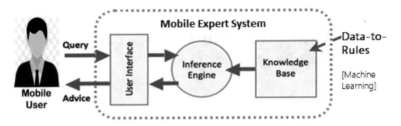

Fig. 8.1 A structure of a mobile expert system modeling

where antecedent represents the conditional statements of a particular situation and the consequent represents the corresponding action. The rule is executed to "fire" when the antecedent is satisfied with the given conditions formulated by contextual information.

- *Inference Engine:* The inference engine, which selects rules from the generated rule-list to execute, is the key processing feature of an expert system. It thus seeks knowledge-based information and relationships and offers answers, or recommendations similar to a human expert. In other words, the knowledge base is used to draw conclusions about the situations. Therefore the main task of the inference engine is to find its way to draw a conclusion through a set of rules. To do this the method of forward chaining can be used, starting from the known facts and moving forward by applying rules of inference to obtain further data, and it continues until it reaches the target. Another method of backward chaining that begins from the goal can also be used according to the needs, going backwards by using inference rules to decide the facts that match the goal.
- *User interface:* The way by which an expert system communicates with a user is referred to as a user interface. Dialog boxes, command prompts, forms, and other input techniques can all be used to do this. Some expert systems interact with other computer applications rather than with humans directly.

8.2   Context-Aware Rule Generation Methods

A context-aware rule has two parts, which follow the "IF-THEN" logical structure to formulate. The antecedent part (premise or condition) represents users' surrounding contextual information such as temporal context, spatial context, social contexts, or others relevant contextual information and the consequent part (conclusion or action) represents their corresponding behavioral activities, or usage [2]. While a rules-based system could be considered as having "fixed" intelligence, in contrast, a machine learning system is adaptive and attempts to simulate human intelligence. There is still a layer of underlying rules, but instead of a human writing a fixed set, the machine can learn new rules independently and discard ones that aren't working anymore. Machine learning techniques can be used to generate such rules from smartphone data. For instance, in our rule mining approach "ABC-RuleMiner", Sarker et al. [3], we have discovered a set of useful contextual rules for mobile phone users considering their behavioral patterns in the data.

Rule-based machine learning approaches include mainly learning classifier systems, e.g., Decision tree [4], and association rule learning, e.g., Apriori algorithm [5] that relies on a set of rules. In Sect. 8.3, we discuss briefly the context-aware rules generated from mobile data. Such an IF-THEN rule-based expert system model can have the decision-making ability of a human expert in an intelligent system designed to solve complex problems and knowledge reasoning. Moreover, the machine learning techniques can be used to update the generated rules according to the recent patterns [6]. In addition to the generated rules, domain experts knowing business rules can also update and manage the rules according to the needs. Thus, mobile expert systems can be used to make intelligent decisions in corresponding mobile applications. In the following, we divide the machine learning rule based methods into two categories such as association rules and classification rules [7, 8] in the field of machine learning and data science, for rule-based modeling.

- *Classification learning rules:* In machine learning, the classification is one of the popular techniques that can be used in various application areas. Several popular classification techniques such as decision trees [4], IntrudTree [9], BehavDT [10], Ripple Down Rule learner (RIDOR) [11], Repeated Incremental Pruning to Produce Error Reduction (RIPPER) [12], etc. exist with the ability of rule generation.

- *Association learning rules:* In general association rules are created by searching for frequent IF-THEN pattern data on the basis of support and confidence value [3]. For generating rules using a given data set, common association rule learning techniques such as AIS [13], Apriori [5], FP-Tree [14], RARM [15], Eclat [16], ABC-RuleMiner [3], etc. can be used. Association rule mining (ARM) techniques are well established in terms of the reliability and accuracy of the rule as has its own parameters [17]. Thus association rules can play a significant role to build a mobile expert system [20].

To extract the context-aware IF-THEN rules using machine learning techniques, we have conducted experiments on two different contextual datasets. One is "phone call dataset" of mobile phone users with different types of calling, e.g. incoming call responses such as answering or rejecting calls, missed and outgoing user calls [18]. The contextual information includes temporal, spatial, and social relationships. To generate rules from the phone call datasets, we have used our earlier behavior-oriented time segmentation (BOTS) technique [19] to pre-process the raw time-series data to create dynamic time segments with similar behavioral patterns. We also generate data-centric social contexts [21] from the raw data for our experimental purpose. Another one is "smartphone apps usage dataset" [22] containing various types of apps usage activities, e.g., using Gmail, Facebook, Youtube, Whatsapp, Browser, Google Maps, etc. in several contexts such as temporal context, work status, spatial context, their emotional state, Internet connectivity, or device-related status. In the following, we discuss how the classification and association rules differ and can be used to build rule-based expert system modeling.

8.3   Context-Aware IF-THEN Rules and Discussion

In this section, we discuss the context-aware IF-THEN rules that are extracted from mobile phone data using the machine learning techniques. We also perform a comparative analysis and discussion between the generated rules considering both the classification and association rules, mentioned in rule-based expert system modeling within the area.

8.3.1   IF-THEN Classification Rules

In this experiment, we first discover the context-aware IF-THEN rules from the phone call dataset mentioned above using the machine learning classification technique. In Table 8.1, we have shown a set of sample generated classification rules considering the temporal context, spatial context, and social relationship, according to the availability in the data. Similarly, in Table 8.2, we have demonstrated another set of generated classification rules utilizing apps usage contextual data, where more number of contexts are involved in the data, mentioned earlier in the dataset descriptions.

If we observe Tables 8.1 and 8.2, we found that the antecedent part (IF) in the rules represents users' contextual information and the consequent part (THEN) represents their corresponding behavioral activities. For instance, the rule R_3 in Table 8.1 states that the user rejects the incoming calls from $Relationship1$ on

Table 8.1 Sample generated classification rules utilizing phone call contextual data

Rules	Context-aware classification rules
R_1	$IF\ time \rightarrow Monday[00:41\ 01:20], relationship \rightarrow Relationship6$ $\Rightarrow THEN\ activity \rightarrow ACCEPT$
R_2	$IF\ time \rightarrow Monday[18:41\ 19:20], relationship \rightarrow Relationship6$ $\Rightarrow THEN\ activity \rightarrow MISSED$
R_3	$IF\ time \rightarrow Tuesday[00:01\ 02:30], relationship \rightarrow Relationship1$ $\Rightarrow THEN\ activity \rightarrow REJECT$
R_4	$IF\ time \rightarrow Tuesday[00:01\ 02:30], relationship \rightarrow Relationship83$ $\Rightarrow THEN\ activity \rightarrow MISSED$
R_5	$IF\ time \rightarrow Saturday[21:16\ 22:30], relationship \rightarrow Relationship39$ $\Rightarrow THEN\ activity \rightarrow OUTGOING$

Table 8.2 Sample generated classification rules utilizing app usage contextual data

Rules	Context-aware classification rules
R_1	$IF\ time \rightarrow Fri[6.00 - 7.00], Location \rightarrow Canteen, Wifi \rightarrow ON$ $\Rightarrow THEN\ activity \rightarrow WebBrowsing$
R_2	$IF\ time \rightarrow Fri[12.00 - 13.00], Location \rightarrow Home, Wifi \rightarrow ON$ $\Rightarrow THEN\ activity \rightarrow UseFacebook$
R_3	$IF\ time \rightarrow Fri[16.00 - 17.00], Location \rightarrow Workplace,$ $Profile \rightarrow Meeting$ $\Rightarrow THEN\ activity \rightarrow UseGmail$
R_4	$IF\ time \rightarrow Sat[14.00 - 15.00], Location \rightarrow Home, Mood \rightarrow Normal$ $\Rightarrow THEN\ activity \rightarrow ReadNews$
R_5	$IF\ time \rightarrow Sun[16.00 - 17.00], Location \rightarrow Home, Mood \rightarrow Normal$ $\Rightarrow THEN\ activity \rightarrow WatchMovie$

Tuesday. Similarly, the rule R_3 in Table 8.2 states that the user uses Gmail at workplace on *Friday* at the *meeting* period.

According to the rules generated in Tables 8.1 and 8.2, we can say that it takes into account the priority of the contexts to generate the rules, e.g., temporal context is common for all the generated rules, due to its highest priority to generate rules. However, the number of contexts in the rules are not static, may vary from rule to rule, depending on the context priority or the impact of the associated contexts to make a decision. Thus its difficult for human experts like the traditional expert system modeling, to assume such priority in contexts to make the expected decisions. The machine learning classification rules that are taken into account in our discussion are capable to handle such context priority in the rules according to their influence in making decisions. Therefore, the context-aware rules considering the impact of the associated contexts of the mobile phone users shown in Tables 8.1 and 8.2 generated from mobile phone data can be used to make the mobile expert system more effective rather than the hardcoded rules.

8.3.2   IF-THEN Association Rules

In this experiment, we first discover the context-aware IF-THEN rules from the phone call dataset mentioned above using the machine learning association technique. In Table 8.3, we have shown a set of sample generated association rules considering the temporal context, spatial context, and social relationship, according to the availability in the data, mentioned above. Similarly, in Table 8.4, we have demonstrated another set of generated association rules utilizing apps usage contextual data, where more number of contexts are involved in the data. As we

Table 8.3 Sample generated association rules utilizing phone call activity contextual data

Rules	Context-aware association rules	Confidence
R_1	$IF\ time \rightarrow Tuesday[17:31\ 18:45], relationship \rightarrow Relationship87,$ $\Rightarrow THEN\ activity \rightarrow ACCEPT$	Conf = 100%
R_2	$IF\ time \rightarrow Tuesday[17:31\ 18:45], location \rightarrow Helsinkicenter,$ $\Rightarrow THEN\ activity \rightarrow MISSED$	Conf = 98%
R_3	$IF\ relationship \rightarrow Relationship91, location \rightarrow Southferry$ $\Rightarrow THEN\ activity \rightarrow OUTGOING$	Conf = 100%
R_4	$IF\ time \rightarrow Sunday[01:16\ 02:30], location \rightarrow Parkslope,$ $relationship \rightarrow Relationship1$ $\Rightarrow THEN\ activity \rightarrow REJECT$	Conf = 92%
R_5	$IF\ time \rightarrow Tuesday[02:31\ 03:45]$ $\Rightarrow THEN\ activity \rightarrow ACCEPT$	Conf = 84%

Table 8.4 Sample generated association rules utilizing app usage contextual data

Rules	Context-aware association rules	Confidence
R_1	$IF\ time \rightarrow Sat[14.00-15.00], ChargingState \rightarrow NotConnected,$ $Mood \rightarrow Happy$ $\Rightarrow THEN\ activity \rightarrow Browsing$	Conf = 100%
R_2	$IF\ location \rightarrow Home, Wifi \rightarrow ON, Mood \rightarrow Sad$ $ChargingState \rightarrow Charging$ $\Rightarrow THEN\ activity \rightarrow WatchYoutube$	Conf = 100%
R_3	$IF\ location \rightarrow OntheWay, Wifi \rightarrow OF, Mood \rightarrow Sad$ $ChargingState \rightarrow Complete$ $\Rightarrow THEN\ activity \rightarrow UseFacebook$	Conf = 91%
R_4	$IF\ holiday \rightarrow Yes, Location \rightarrow Home, Wifi \rightarrow ON, Mood \rightarrow Sad$ $\Rightarrow THEN\ activity \rightarrow WatchMovie$	Conf = 88%
R_5	$IF\ time \rightarrow Sat[6.00-7.00], Holiday \rightarrow Yes, Location \rightarrow Home,$ $Wifi \rightarrow OFF, Mood \rightarrow Happy$ $\Rightarrow THEN\ activity \rightarrow ReadNews$	Conf = 80%

produce mobile user behavioral association rules for a given confidence preference, the results are presented in Tables 8.3 and 8.4, when the preferred confidence level is, say, 80%.

If we observe Tables 8.3 and 8.4, we found that the antecedent part (IF) in the rules represents users' contextual information and the consequent part (THEN) represents their corresponding behavioral activities like the classification rules discussed above. However, the association rules have their corresponding confidence values that measure the strength of the rules, shown in Tables 8.3 and 8.4. For instance, the rule R_4 in Table 8.3 states that the user rejects most of the incoming calls (92%) from *Relationship*1 on *Sunday*, when she is at *Parkslope*. Similarly, the rule R_4 in Table 8.4 states that the user watches Movie at *home* (88%), when she is in *sadmood* and *WiFiconnectivity* is on.

According to the rules generated in Tables 8.3 and 8.4, we can say that the number of contexts in the rules are not static, may vary from rule to rule, depending on the associated contextual patterns. For some rules, only a few number of contexts or even one context, e.g., rule R_5 with only temporal context in Table 8.3, can make the decisions with high confidence value, and in some cases, the rules consist of higher dimensions of contexts, e.g., rule R_5 consists of temporal, work day status, spatial, Internet connectivity, and user mood in Table 8.4. Thus its difficult for human experts like the traditional expert system modeling, to assume such dynamism in context-aware rules to make the expected decisions according to personalized preferences. Moreover, the preferences in association rules may vary from user to user in the real world life and diverse situations. The machine learning association rules that are taken into account in our discussion are capable to handle such dynamism according to the data patterns, which can effectively make the decisions based on the associated contexts. As the association rules represent the confidence values, these rules are more reliable than the above generated classification rules, to make decisions according to users' preferences. Therefore, the behavior-oriented rules of mobile phone users with the confidence values shown in Tables 8.3 and 8.4 generated from mobile phone data can be used to make the mobile expert system more effective.

8.4 Conclusion

In this chapter, we have discussed the context-aware rule-based expert system modeling for mobile data science applications to make intelligent decisions. The rule-based expert system is considered one of the key artificial intelligence techniques that can be used to make intelligent and more powerful applications in the area. Our analysis of context-aware rule-based expert system modeling can play a significant role in designing and building data-driven intelligent mobile systems. The rule based expert systems can provide various personalized smartphone services in the real-world life. Such systems, either standalone or distributed, may assist the users intelligently according to the generated rules in different contextual day-to-day situations in their daily life. Overall, our study on rule-based expert system modeling will aid application developers in creating context-aware intelligent applications that will intelligently support smartphone users in their daily tasks. In addition to mobile applications, the concept of our expert system modeling can also be applicable in different application domains, e.g., in the context of cybersecurity, this can help to build machine learning rule-based intelligent security systems by taking into account the generated policy rules as knowledge-base.

References

1. Sarker, I. H., Hoque, M. M., Uddin, M. K., & Alsanoosy, T. (2021). Mobile data science and intelligent apps: concepts, AI-based modeling and research directions. *Mobile Networks and Applications, 26*(1), 285–303.
2. Sarker, I. H. (2019). Context-aware rule learning from smartphone data: survey, challenges and future directions. *Journal of Big Data, 6*(1), 1–25.
3. Sarker, I. H., & Kayes, A. S. M. (2020). ABC-RuleMiner: User behavioral rule-based machine learning method for context-aware intelligent services. *Journal of Network and Computer Applications, 168*, 102762.
4. Quinlan, J. R. (2014). *C4. 5: programs for machine learning.* Elsevier.
5. Agrawal, R., & Srikant, R. (1994, September). Fast algorithms for mining association rules. In *Proc. 20th Int. Conf. Very Large Data Bases, VLDB* (Vol. 1215, pp. 487–499).
6. Sarker, I. H., Colman, A., & Han, J. (2019). Recencyminer: mining recency-based personalized behavior from contextual smartphone data. *Journal of Big Data, 6*(1), 1–21.
7. Han, J., Kamber, M., & Pei, J. (2011). Data mining concepts and techniques third edition. *The Morgan Kaufmann Series in Data Management Systems, 5*(4), 83–124.
8. Sarker, I. H. (2021). Machine learning: Algorithms, real-world applications and research directions. SN Computer Science, 2(3), 1–21.
9. Sarker, I. H., Abushark, Y. B., Alsolami, F., & Khan, A. I. (2020). Intrudtree: a machine learning based cyber security intrusion detection model. *Symmetry, 12*(5), 754.
10. Sarker, I. H., Colman, A., Han, J., Khan, A. I., Abushark, Y. B., & Salah, K. (2020). Behavdt: a behavioral decision tree learning to build user-centric context-aware predictive model. *Mobile Networks and Applications, 25*(3), 1151–1161.
11. Witten, I. H., Frank, E., Hall, M. A., & Pal, C. J. (2005). *Practical machine learning tools and techniques* (p. 578). Morgan Kaufmann.
12. Witten, I. H., Frank, E., Trigg, L. E., Hall, M. A., Holmes, G., & Cunningham, S. J. (1999). Weka: Practical machine learning tools and techniques with Java implementations.
13. Agrawal, R., Imieliński, T., & Swami, A. (1993, June). Mining association rules between sets of items in large databases. In *Proceedings of the 1993 ACM SIGMOD International Conference on Management of Data* (pp. 207–216).
14. Han, J., Pei, J., & Yin, Y. (2000). Mining frequent patterns without candidate generation. *ACM sigmod Record, 29*(2), 1–12.
15. Das, A., Ng, W. K., & Woon, Y. K. (2001, October). Rapid association rule mining. In *Proceedings of the Tenth International Conference on Information and Knowledge Management* (pp. 474–481).
16. Zaki, M. J. (2000). Scalable algorithms for association mining. *IEEE Transactions on Knowledge and Data Engineering, 12*(3), 372–390.
17. Freitas, A. A. (2000). Understanding the crucial differences between classification and discovery of association rules: a position paper. *ACM sIGKDD Explorations Newsletter, 2*(1), 65–69.
18. Eagle, N., & Pentland, A. S. (2006). Reality mining: sensing complex social systems. *Personal and Ubiquitous Computing, 10*(4), 255–268.
19. Sarker, I. H., Colman, A., Kabir, M. A., & Han, J. (2018). Individualized time-series segmentation for mining mobile phone user behavior. *The Computer Journal, 61*(3), 349–368.
20. Sarker, I. H., Khan, A. I., Abushark, Y. B., & Alsolami, F. (2021). Mobile Expert System: Exploring Context-Aware Machine Learning Rules for Personalized Decision-Making in Mobile Applications. *Symmetry, 13*(10), 1975.
21. Sarker, I. (2018). Understanding the role of data-centric social context in personalized mobile applications. *EAI Endorsed Transactions on Context-Aware Systems and Applications, 5*(15), 1–6.
22. Sarker, I. H., & Salah, K. (2019). Appspred: predicting context-aware smartphone apps using random forest learning. *Internet of Things, 8*, 100106.

Chapter 9
Deep Learning for Contextual Mobile Data Analytics

9.1 Introduction

In the field of data analytics, various machine learning techniques, such as Decision Trees, Random Forests, Support Vector Machines, Logistic Regression, Adaptive Boosting are common and can be used for building context-aware prediction model [1]. Real-world mobile usage data, on the other hand, may include many contextual dimensions and be much larger in size due to the users' daily behavioral data. As a result, conventional machine learning models may not be suitable for developing a context-aware model. While principal components analysis (PCA) [2], an unsupervised, non-parametric statistical technique in machine learning, is an effective tool for reducing dimensions, it may often lose important information when transforming contextual attributes to components.

Deep learning [3], on the other hand, is a subset of machine learning algorithms or artificial intelligence that evolved from the artificial neural network (ANN), as illustrated in Fig. 9.1, which employs multiple layers to extract higher-level contextual features from raw data. Deep learning differs from conventional machine learning in terms of efficiency as the volume of data increases. Furthermore, deep learning and conventional machine-learning algorithms vary significantly in their ability to extract high-level features directly from data. While deep learning takes a long time to train a model due to the large number of parameters, it takes a short amount of time to run during testing as compared to other machine learning algorithms.

Deep learning is progressing due to the availability of powerful computational resources and large amounts of training data. Computation on mobile devices is becoming possible as mobile devices become more computationally efficient [4, 5]. If deep learning in the cloud is suitable for the application, using one of many current cloud artificial intelligence APIs could be the simplest way to deploy deep learning capabilities on a mobile device. In this case, the device acts as both a sensor and a user interface. These APIs include AI capabilities in machine learning, speech

© The Author(s), under exclusive license to Springer Nature Switzerland AG 2021
I. H. Sarker et al., *Context-Aware Machine Learning and Mobile Data Analytics*,
https://doi.org/10.1007/978-3-030-88530-4_9

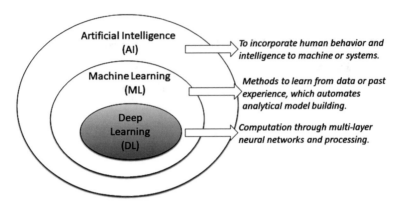

Fig. 9.1 Deep learning within the area of machine learning and artificial intelligence

recognition, natural language processing, AI assistant, computer vision, and other areas. Microsoft Cognitive Services, Google Cloud Vision, IBM Watson Services, Amazon are some examples of well-known services [4]. This chapter discusses the importance of deep learning in context-aware behavior modeling for mobile phone users and explore a deep neural network learning-based context-aware model.

Popular deep learning techniques, such as Multi-layer Perceptron (MLP), Convolutional Neural Network (CNN or ConvNet), Recurrent Neural Network (RNN) or Long Short-Term Memory (LSTM), Self-organizing Map (SOM), Auto-Encoder (AE), Restricted Boltzmann Machine (RBM), Deep Belief Networks (DBN), Generative Adversarial Network (GAN), Deep Transfer Learning (DTL or Deep TL), Deep Reinforcement Learning (DRL or Deep RL), or their ensembles and hybrid approaches can be used to solve problems in various application domains [3, 6]. A typical artificial neural network model, i.e., MLP is a completely connected network that includes an input layer that receives input data, an output layer that makes a decision or prediction about the input signal, and one or more hidden layers between these layers, which are called the network's true computational engine, as shown in Fig. 9.2. In this chapter, we primarily focus on context-aware smartphone use prediction using such deep learning modeling with multiple hidden layers.

9.2 Contextual Data

Data availability is typically the basis for a data-driven model based on ANN and DL methods [1]. Typically, datasets represent a collection of data records with a variety of attributes or characteristics, as well as other relevant information, from which the data-driven model is derived. Several contextual data are considered in this study, including not only the user-centric context, such as the user's spatio-temporal context, mood or emotional state, and so on, but also the device-centric context, such

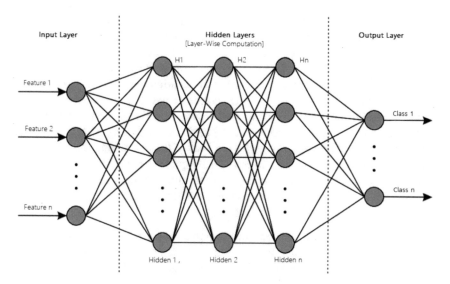

Fig. 9.2 An example of neural network with multiple hidden layers

as battery level, phone profile, Internet access, and so on that are discussed below. Different types of mobile apps are considered, including social networking, instant messaging, voice communications, entertainment, and other apps that are relevant to users' everyday lives.

- *User-Centric Contexts:* According to the general concept of context, it could be anything to define the state of an object or entity [5]. When establishing the context in this work, a smartphone user can be interpreted as an entity. As a result, different data measurements can have an effect on how mobile users use their apps. For example, temporal context reflects time-related information on how users interact with apps. It is one of the most important situations that has a significant impact on mobile users' phone behaviors [7]. In addition to such temporal data, users' employment status could be another factor that has a major impact on app use for many people. For example, an individual user's app usage behavior on Saturday, say a holiday, can vary from her usage on Monday, the first working day of the week [8]. While it is comparable to the temporal sense in terms of weekdays and weekends, it also represents individuals' working status, which is an important context for modeling mobile app usage based on their preferences. A spatial context that reflects user spatial knowledge, such as one of the current office locations, is another relevant context to consider [5]. For the purpose of developing human-centered context-aware application, the spatio-temporal context is popular. User mood or emotional state can be another important context that influences people, especially in human-centered applications. When a person is in a good mood, for example, she prefers to listen to only her favorite songs, while she prefers to talk with her close friends on social media, when she is sad. We believe that all of these factors will have an impact on predicting how context-aware smartphones will be used, and could vary from user to user depending on their preferences.

- *Device-Centric Contexts:* Users' device-centric contexts are also relevant for modeling the behavior of user app usage, in addition to the above user-centered contexts related to day-to-day circumstances or personal preferences of users. Contextual information can include one's contact profile, phone's battery level or charging status, Internet access, and so on, all of which may have influence on how different mobile app categories are used [5]. For example, if one's device has a low-power signal, she will not usually connect her device to the Internet in order to use an entertainment app like watching Youtube videos. As a result, all of these device-centric contextual data could also play an important role on predicting context-aware smartphone usage.

To predict the usage of context-aware mobile apps, we use both user-centric and device-centric contextual information in our deep learning modeling. We summarized the detailed picture of the contexts that are used in our deep learning model in Table 9.1. Figures 9.3 and 9.4, for example, show two separate function data distributions, *time* and *workday* respectively. The value is very small for some data points, while it is much higher for others, as seen in Figs. 9.3 and 9.4. We will need exploratory data analysis to feed our target artificial neural network learning classification technique, as discussed above, in order to build our deep learning model. Missing data handling, exploring contextual feature encoding methods like label encoding or one hot encoding, feature scaling, normalization of the contextual data etc. are common while building a data-driven model [5].

Table 9.1 An overview of contexts used in our deep learning model

Contexts	Type	Example values
Temporal context	Continuous	Time-of-the-day [24-hours-a-day] Days-of-the-week [7-days-a-week]
Work status context	Categorical (binary)	Workday and Holiday
Spatial context	Categorical	Phone user location [at home, at office, at the canteen, in the playground, on the way, etc.]
User mood context	Categorical	Emotional state of phone user [normal, happy, or sad]
Phone profile context	Categorical	Phone notification [general, silent, or vibration]
Battery charging status context	Categorical	Battery level [low, medium, or full]
Internet connectivity context	Categorical (binary)	WiFi connectivity [on, off]
Smartphone apps	Categorical	Social networking, Gmail, Communication, Video, Entertainment, Read News, Games etc.

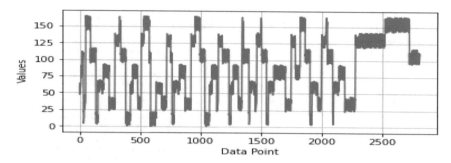

Fig. 9.3 Data distribution of the contextual feature '*time*'

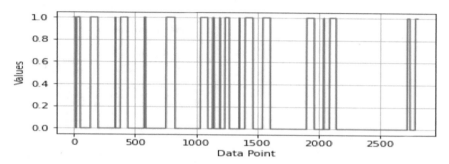

Fig. 9.4 Data distribution of the contextual feature '*work day*'

9.3   Deep Neural Network Modeling

In this section, we present a context-aware prediction model for smartphone users based on deep neural network learning.

9.3.1   Model Overview

As shown in Fig. 9.5, the model is built with multiple layers. The neurons in the network are also known as nodes, and they are linked together in different layers by connections. We use the contextual features selected through the correlation analysis as the size of the input layer, and an output layer with the number of neurons equal to the number of app classes, i.e., a multi-class classification task, to construct our context-aware model. For computation, we use multiple hidden layers with up to 400 neurons. We also use dropout in each layer to simplify the model and use the Adam optimizer to compile the neural network model. When training the contextual network, we use 500 epochs with a batch size of 128. We often use a tiny 0.001 learning rate since it allows the contextual network model to reach the global minimum. To adjust model weights, the loss function defined in Eq. 9.3 is used. We

Fig. 9.5 Our deep neural
network learning based
context-aware smartphone
apps usage prediction model

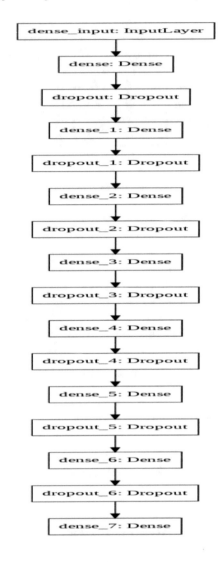

empirically set these hyperparameters to build our deep learning model using deep
learning from artificial neural networks. In the following, we will go through each
layer of our model separately.

9.3.2   Input Layer

The contextual data is transferred directly from the model input layer to the first
hidden layer, where the data is multiplied by the weights of the first hidden layer.
The input contextual data is chosen using the correlation analysis, and the size of
the input layer is determined by the number of contextual features chosen. We do

this filtering to eliminate the less important, or redundant context from the given dataset, by analyzing the data patterns and dependency, in order to create an efficient context-aware model. This filtering reduces the size of the input layer's neurons and, as a result simplifies the model. While building the deep learning model, we consider contextual features in the input layer, such as temporal context, work status (weekday or holiday), spatial context, emotional state, Internet access, phone profile, and system battery level, based on their relevance to the target class.

9.3.3   Hidden Layer(s)

The nodes of this layer are not visible to the outside world, but they are part of any neural network's abstraction. The hidden layer takes measurements of all the features entered through the input layer and sends the results to the output layer. In a nutshell, the hidden layers perform nonlinear transformations on the inputs to the network. As the activation function, we use the Rectified Linear Unit (ReLU) defined in Eq. 9.1. ReLU solves the vanishing gradient problem while also allowing the model to learn faster [9, 10]. We often use dropouts in each layer to simplify the model and use the "Adam" optimizer to compile the neural network model. Adam optimization is a stochastic gradient descent method for adaptive first and second-order moment estimation. To adjust the weights of the model, we use the *Cross Entropy* loss function, which is defined in Eq. 9.3. During learning, the most common Backpropagation process [11], is used to update the model's relation weights between neurons.

$$ReLU : f(x) = max(0, x) \tag{9.1}$$

$$Softmax : f(y_k) = \frac{\exp(\phi_k)}{\sum_j^c \exp(\phi_j)} \tag{9.2}$$

$$Loss = -\sum_{i=1}^{n}\sum_{j=1}^{m} y_{i,j} \log(p_{i,j}) \tag{9.3}$$

Where, $y_{i,j}$ denotes the true value, i.e., 1 if the i sample belongs to the j and 0 class otherwise and $p_{i,j}$ denotes the likelihood of the i sample model belonging to the j class.

9.3.4   Output Layer

This layer communicates the network's information to the outside world. As a result, this output layer is in charge of producing the final prediction's outcome. The output layer takes the inputs from the layers that come before it, runs the calculations through its neurons, and then computes the output. The number of neurons of this layer corresponds to the size of various apps such as Facebook (FB), Gmail, Movie (MOV), Skype (SK), Music (MS), LinkedIn (LI), Live Sport (LS), Whatsapp (WA), Browsing (BR), Read News (RN), Instagram (IG), Youtube (YT), Games (GM), and so on. As a result, our model is considered as a multi-class classification task, with these different app categories serving as model classes. In our deep learning model, we use the 'Softmax' activation function described in Eq. 9.2, which produces values between 0 and 1 for each of the outputs that add up to 1. As a result, this can be deduced as a multinomial probability distribution. To accomplish our task, the softmax function is used as the activation function in the output layer of neural network models.

9.4 Prediction Results of the Model

This experiment shows the prediction results of our deep neural network learning-based model. The calculated outcome for predicting context-aware smartphone usage is shown in Figs. 9.6 and 9.7. The results are shown in terms of model accuracy and loss score for user U1 and U2 respectively.

Figures 9.6 and 9.7 show that the accuracy is initially low and the loss score is high. However, as the number of epochs increases, the accuracy improves while the loss score decreases. The reason for this is that our model starts with random weight values to build the network.

It creates the correct model by factoring in weight updates at each epoch. At the end, it will be able to build a context-aware app usage model with a higher accuracy score and a lower loss score for the unknown test cases. As a result, we can deduce that the mobile deep learning model can predict smartphone usage based on contextual information with higher accuracy and lower model loss, as shown in Figs. 9.6 and 9.7.

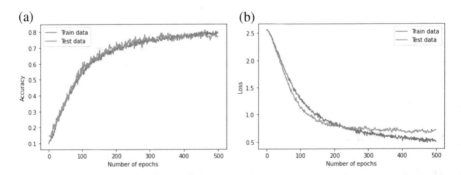

Fig. 9.6 Calculated outcome in terms of accuracy and loss score of the mobile deep learning model for predicting context-aware smartphone usage utilizing the dataset of User $U1$. (**a**) Model accuracy score. (**b**) Model loss score

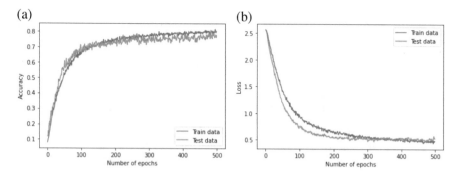

Fig. 9.7 Calculated outcome in terms of accuracy and loss score of the mobile deep learning model for predicting context-aware smartphone usage utilizing the dataset of User $U2$. (**a**) Model accuracy score. (**b**) Model loss score

9.5   Conclusion

We present a deep neural network learning model for predicting context-aware smartphone usage in this paper. In our model, we took into account contextual information in a variety of dimensions, including temporal context, such as workday or holiday status, spatial context, user emotional states, WiFi status or Internet connection, and device-related status, such as charging status, profiling, and so on. In order to build the model, we took into account a variety of contextual features that could have influence on how people use apps in their various real-world contexts on a daily basis, as well as data sets on app usage from smartphone users. The results of the experiments on the usage of datasets for mobile apps show that the model is capable of building an effective context-aware prediction model. We believe that this model would be useful for application developers to build suitable real-life applications for end-users, particularly where conventional machine learning models struggle with higher context dimensions and large amounts of contextual data.

References

1. Sarker, I.H. (2021). Data Science and Analytics: An Overview from Data-Driven Smart Computing, Decision-Making and Applications Perspective. *SN Computer Science, 2*(377), 1–22.
2. Sarker, I. H. (2021). Machine learning: Algorithms, real-world applications and research directions. *SN Computer Science, 2*(3), 1–21.
3. Sarker, I. H. (2021). Deep Learning: A Comprehensive Overview on Techniques, Taxonomy, Applications and Research Directions. *SN Computer Science, 2*(6), 1–20.
4. Deng, Y. (2019, May). Deep learning on mobile devices: a review. In *Mobile Multimedia/Image Processing, Security, and Applications 2019* (Vol. 10993, p. 109930A). International Society for Optics and Photonics.

5. Sarker, I. H., Abushark, Y. B., Khan, A. I., Alam, M. M., & Nowrozy, R. (2021). Mobile deep learning: Exploring deep neural network for predicting context-aware smartphone usage. *SN Computer Science, 2*(3), 1–12.

6. Sarker, I. H., Furhad, M. H., & Nowrozy, R. (2021). AI-driven cybersecurity: An overview, security intelligence modeling and research directions. *SN Computer Science, 2*(3), 1–18.

7. Sarker, I. H., Colman, A., Kabir, M. A., & Han, J. (2018). Individualized time-series segmentation for mining mobile phone user behavior. *The Computer Journal, 61*(3), 349–368.

8. Sarker, I. H., & Salah, K. (2019). Appspred: predicting context-aware smartphone apps using random forest learning. *Internet of Things, 8*, 100106.

9. Géron, A. (2019). *Hands-on machine learning with Scikit-Learn, Keras, and TensorFlow: Concepts, tools, and techniques to build intelligent systems.* O'Reilly Media.

10. Cicceri, G., De Vita, F., Bruneo, D., Merlino, G., & Puliafito, A. (2020). A deep learning approach for pressure ulcer prevention using wearable computing. *Human-centric Computing and Information Sciences, 10*(1), 5.

11. Han, J., Kamber, M., & Pei, J. (2011). Data mining concepts and techniques third edition. *The Morgan Kaufmann Series in Data Management Systems, 5*(4), 83–124.

Chapter 10
Context-Aware Machine Learning System: Applications and Challenging Issues

10.1 Rule-Based Intelligent Mobile Applications

A machine learning rule-based system expresses information as a collection of IF-THEN rules extracted from the data using machine learning techniques [1]. This rule-based system can specify what to do or infer in various scenarios and can act as a software agent [2]. A real-life smartphone application is an actual platform to employ the discovered contextual rules. The development of individual real-life applications that integrate the behavioral rules to make rule-based predictions or to provide the personalized services for individual mobile phone users in a context-aware pervasive computing environment. For example, an "intelligent phone call interruption management system" could be a real-world application for mobile users based on the discovered behavioral rules that automatically manage interruptions according to their preferences. Similarly, discovered rules from mobile apps logs can be used to build a "mobile recommendation system" to assist them in their daily life according to the needs of an individual in different contexts. In the following, we have discussed several applications considering context-awareness that have been studied widely in the past few years.

- *Smart Context-Aware Mobile Communication*: Although mobile phones are considered to be "always on, always connected" devices, users are not always aware of and responsive to incoming mobile communication [3]. As a result, people are often disrupted by incoming phone calls in different day-to-day circumstances in their everyday lives, causing disruption not just to the phone users but also to others nearby. Such interruptions can cause embarrassment not only in a professional environment, such as a meeting or a lecture but also in other activities like observing patients by a doctor or driving a car. These types of interruptions may often lead to decreased worker efficiency, increased errors, and stress in the workplace [4]. Call activity records in application logs (e.g., phone call logs) are a rich resource in the real world for mining contextual behavioral rules of individual cell phone users, which can be used for building smart call interruption management systems to intelligently handling incoming calls according to their preferences [5]. Another real-world application is a smart call reminder system that intelligently finds the desired contact from a huge contact list and reminds a user to make a phone call to a specific person in a specific context, based on the user's behavioral rules discovered from their previous calling history.

- *Intelligent Mobile Notification Management*: These days one can find a wide range of smart smartphone apps from the app stores. These applications enable users of mobile phones to subscribe to a large number of information channels and actively receive a large number of notifications [6]. However, depending on the content form and sender of the messages, cell phone users do not approve these notifications. Users usually ignore notifications that are not useful or important to their interests [6]. Most notifications of inviting games on social networks, social or promotional emails, for example, are swiped away without being clicked, indicating that the user has no interest in these notifications. Furthermore, predictive recommendations from various cell phone apps, such as social networking systems, e.g., Facebook, WhatsApp, Viber, Skype, and Youtube, may or may not be of interest to a specific user [6–8]. The explanation for this is that users may become annoyed by such uninteresting phone notifications. As a result, some users choose to uninstall the relevant apps from their smartphones to stop receiving such updates. Individual behavioral rules based on contextual information can be used to intelligently handle such notifications. For instance, one person often ignores promotional email notifications; one accepts Facebook birthday reminder notifications mostly at night when she is at home; and one does not accept Viber or Whatsapp notifications from unknown people at work. Such behavioral activities may vary from user to user depending on their preferences in various contexts. The contextual behavioral rules can be used to provide such personalized services intelligently to mobile phone users.
- *Context-Aware Mobile Recommendation*: Traditional recommender systems [9] primarily concentrate on recommending the most appropriate products or services to users. When making a recommendation to a specific person, these traditional recommendation systems typically do not take into account the contextual information [10]. However, in certain real-life scenarios, it may not be sufficient to make recommendations without taking into account such contextual information. For example, depending on the temporal context, the recommendation outcome of a travel recommender system in the summer can be very different from the outcome in the winter for a specific user. Similarly, a particular context, such as location information, can have an impact on making different recommendations for users [11, 12]. One of the most important aspects of mobile recommendation is mobile app recommendation [13]. Mobile devices such as smartphones and tablets have become one of the most significant media for social entertainment and knowledge acquisition due to their rapid growth and adoption [14]. In reality, various contexts and app usages (e.g., Multimedia, Facebook, Gmail, Youtube, Skype, Game) data is captured in context-rich system

logs, which can be used to mine the contextual behavioral rules of individual mobile phone users, i.e., which app is preferred by a specific user in a given context. Mining such preferences, in particular, is a crucial step in gaining a better understanding of mobile phone users' app use habits. The context-logs-based behavioral rules can be used to provide customized context-aware recommendations for various mobile phone applications to mobile phone users.

- *Smart Context-Aware Mobile Tourist Guide*: Tourists today expect customized access to tourism information at any time, from any place, and through any medium. Traditional information centers provide tourists with valuable and applicable information, but they are not available at any time or in any place. As circumstances or context parameters change, tourists may wish to change their schedules or re-plan their itineraries. As a result, a ubiquitous and pervasive tourist assistant can play a crucial role in the tourism industry's growth. A context-aware web service-based tourism information system can be a smart tourist guide or information system. If context elements are applied to these systems, they will be able to present more specific information based on the user's needs and the context, for example, his or her current location or time. Consider the following scenario: the system demands restaurant details around the tourist's current venue, say, Melbourne, Australia, at noon. The information collected takes into account the tourist's food preferences as well as his or her economic status. Finally, the tourist's mobile device displays the appropriate restaurant list. In this task, the elements of context are tourist's location (e.g., Melbourne, Australia), the time of day (e.g., noon), and personal interests (may differ from user-to-user). Such a context-aware mobile application allows travel enjoyable and provides useful information without requiring the users to expend too much time or money. Therefore, a rule-based context-aware system can help to generate a personalized guide for the tourists based on history.
- *Rule-based Predictive and Personalized Services*: Predictive modeling, in general, employs historical data or statistics to forecast a relevant future outcome that can be applied to any uncertain occurrence, regardless of when it happened. As a result, individual mobile phone users' contextual behavioral rules can be used to predict their actions in response to specific contextual information. Some examples of such predictions are—to predict the outgoing calls analyzing mobile phone historical call log data [15–17] for smart searching in contact list, to predict incoming calls for planning and scheduling (e.g., it can be used to avoid unwanted calls and schedule time for wanted calls) [18], to predict the next mobile application that an individual is going to use for a particular contexts by analyzing individual's app usages data [19–22], to predict smartphone notification response behavior of individual users utilizing their responses to the notifications stored in the smartphone notification logs, in order to build intelligent notification management system [7, 8]. In addition to these mobile usage related services, rule-based modeling can be used in other personalized services like smart-city services, health services, transport services, etc. to assist them in their daily activities in different situations in a context-aware pervasive computing environment.

- *Context-Aware Self Management and Energy Saving*: Due to the limited energy supply, energy conservation is a crucial problem for real-time systems in embedded devices. In most situations, the battery is the primary source of power for smart mobile devices. Due to the number of processing resources available in these devices, users can now accomplish a variety of tasks that were previously only possible with a computer. These devices, however, continue to have problems with power management. As smartphones are equipped with many features, users need to manage a growing number of complex configuration settings, which allows automated configuration of the devices without taking attention to users. Context-aware smart configuration settings including volume settings, WiFi turn on/off, GPS turn on/off, and application management can play a role to reduce battery consumption. Users desire different groups of settings to be applied for different contexts. However, this is currently difficult to do, because it is typically a manually-intensive process that cannot adapt to changing context. A smart self-management system automatically changes the smartphone configuration when the smartphone needs to be changed and what configuration settings it needs to have for a given context.
- *Context-Aware Smartcity Services*: A smart sustainable city is an innovative city that uses information and communication technologies (ICTs) and other means to improve quality of life, efficiency of urban operations and services, and competitiveness while also meeting current and future generations' economic, social, and environmental needs. There are variety of smartcity applications ranging from personalized to population services. For instance, a rule-based tourist recommendation system could be a smartcity application [23]. Similarly, traffic light management, smart grid, smart building and home services, smart parking, etc. are popular smartcity applications. The extracted rules based on contexts from the relevant data source can play a significant role to build the corresponding model to provide these services more intelligently.
- *Context-Aware Security and Privacy*: Security ensures the confidentiality, integrity, and availability of information in general, whereas privacy is more specific to personal information privacy rights [24]. Because of the threats associated with IT consumerization and cloud computing, context-aware security has become more relevant in recent years. This is a practical method for implementing user-centric security and privacy since it allows for the management of threat models associated with the users' frequent context changes. A security context is a collection of information gathered from the user's environment and the application environment that is relevant to the security infrastructure of both the user and the application domain. As a result, context-aware security refers to the use of contextual data to improve information security decisions such as identity, geolocation, time of day, and endpoint device type. The majority of current research at the application layer focuses on context-based security policies for adaptive authentication and authorization services [25]. Adaptive security policies are described by the International Telecommunication Union (ITU) as a collection of security actions about different layers of the security architecture: Infrastructure layer (physical network

nodes and communication links), service layer (basic networking, transport, and added value services), and application layer (network-based applications) [25, 26]. Thus context-aware rule-based security policies can play a vital role in the domain of mobile platforms as well as Internet of Things (IoT) applications.

Overall, the contextual rule-based modeling can be used in various real-world application areas, where the context-awareness and application intelligence are involved. Thus, the impact of contextual rule learning in mobile application development and user experience is significant in these days and can be considered as next-generation mobile learning.

10.2 Major Challenges and Research Issues

We summarize a range of research issues relating to mining contextual behavioral rules of individual cell phone users in this section. These include mobile data quality, determining the relevance of contexts to provide dynamic services, discretization of continuous contextual data as the foundation for knowledge discovery, user behavioral rule discovery and ranking, knowledge-based interactive post-mining for semantic comprehension, and dynamic updating and management of rules based on their current behavior. In the following, we briefly discuss these issues.

- *Collection and Management of Contextual Data:* Collecting real-world contextual data is the first step to build data-driven intelligent applications for mobile phone users [27]. The reason is that such data usually comprises features whose interpretation depends on some contextual information, such as temporal, spatial, or social context, relevant to mobile phone users. The contextual data can be acquired from distributed sources. For instance, users' social activity such as 'in a meeting' and relevant information can be acquired from the calendar, and users' mobile phone usage information and corresponding contextual information can be acquired from different sensors or context sources, such as phone logs [28]. Thus, to facilitate the extraction of reliable insight from contextual information and to use the knowledge in context-aware intelligent applications, integrating and effective management of relevant contexts is important. Therefore, the challenge is to acquire contextual information from distributed sources and how to integrate and manage such information for effective data analysis. Developing an incentive program for mobile devices can be used for collecting timely, complete, consistent, and accurate data effectively.
- *Ensuring the Quality of Smartphone Data:* Since mobile phone data is collected and stored using a variety of sensors and data sources, it may contain noise, i.e., wrong and/or redundant instances. Noise is the term for such inconsistency in a mobile phone dataset. Simply stated, noise is something that obscures the relation between an instance's features or contexts and its behavior class in a dataset [29]. The existence of such noisy instances in cell phone data is a major problem for modeling user behavior, with numerous adverse consequences. For example,

due to the number of incorrect or redundant training samples, the over-fitting problem can occur, lowering prediction accuracy and increasing the complexity of machine learning techniques [30].

As the prediction model requires noise-free training data collection, it's difficult to get higher classification or prediction accuracy of user activity by analyzing individuals' mobile phone log data using machine learning techniques. According to Sarker et al. [31], the effects of noisy instances are: noise can result in the creation of additional behavioral rules that are irrelevant to individual cell phone users, resulting in an unnecessarily large ruleset; the training samples may increase, thus increasing the complexity of the corresponding machine learning-based behavioral model for individuals. To ensure the accuracy of the training data, it is essential to identify and remove the noisy instances from the dataset. The consistency of the training data and the competence of the machine learning algorithm are both important factors in the performance of the machine learning technique-based model. As a result, before constructing the model, a noise reduction process is needed to improve the model's prediction accuracy based on real-world cell phone data.

- *Understanding the relevancy of contexts:* Realizing the importance of contexts is a crucial step towards effectively using them in mining contextual behavioral rules of individual cell phone users. We need a better understanding of what circumstances influence users to make decisions in various situations to effectively use contexts in the behavioral rules of individual cell phone users. The contexts related to the user are the most applicable as we try to discover the contextual behavioral rules of individuals using their cell phone data. However, the contexts' relevancy is application-specific, i.e., it may differ from one application to another in the real world.

Consider a customized smart mobile app management system that can predict an individual's future app usages (e.g., Skype, Whatsapp, Facebook, Gmail, Microsoft Outlook, and so on) based on contextual data. When the user is at her office on weekdays between 9:00 a.m. and 10:00 a.m., she usually uses Microsoft Outlook for mailing purposes. The user's contexts, such as temporal (weekdays between 09:00 AM and 10:00 AM) and location (at the office), may be important to intelligently assist herself in finding this specific mobile application among a large number of installed applications on her mobile phone. Consider another scenario, such as a mobile phone call interruption management system, where more contexts might be relevant. Cell phones are often referred to as "always-on, always-connected" devices in the real world, but mobile users are not always attentive and responsive to incoming communication [3]. Let's say a user has a routine meeting at her office on Monday between 9:00 and 11:00 a.m. She usually declines incoming phone calls during that period because she does not want to be distracted during the meeting. If the phone call is from her boss or mother, she wants to answer it because it seems to be important to her. According to this example, user phone call answer patterns are relevant not only to the above contexts, location (e.g., at the office), and temporal (e.g., on Monday, between

9:00 AM and 11:00 AM), but also to the additional contexts, social situation (e.g., in a meeting), and social-relational context (e.g., boss or mother).

The relevancy of user contexts differs from application to application in the real world, as shown by the real-world examples above. As a result, a greater understanding of context relevancy as it relates to individual users' needs would aid mobile app developers in deciding what context to use in their apps to offer customized services that assist them intelligibly.

- *Discretization of Contextual Data:* Discretization is one of the most important preprocessing techniques in data mining since it functions as the basis for identifying useful knowledge or rules [27]. Depending on such parameters, the discretization process converts continuous numerical attribute values into discrete or nominal attribute values. To put it another way, it converts quantitative data into qualitative data with a finite number of intervals, resulting in a non-overlapping partition in a continuous domain such as time or location. The essence of continuous data is that it is always massive in size, has high dimensionality, and is updated regularly. Assuming a data set consisting of N samples and C target classes, a discretization algorithm would discretize the continuous attribute A in this data set into m discrete intervals, $Dis = [d_0, d_1]$, $[d_1, d_2], ..., [d_{m-1}, d_m]$, where d_0 represents the minimal value, d_m represents the maximal value, and $d_i < d_{i+1}$, for $i = 0, 1, .., m - 1$. Such a discrete result Dis is called a discretization scheme on attribute A, and $< p = d_1, d_2, .., d_{m-1} >$ is the set of cut points of attribute A.

  Let's consider, time-series data, which is the most important continuous context that impacts user behavior in a mobile Internet portal [32]. Mobile phones keep track of the precise temporal details (e.g., YYYY-MM-DD hh:mm:ss) of users' activities with their phones. In contrast to digital systems, human interpretation of time is not precise in behavior modeling. Every routine activity requires a period, such as five minutes. For example, a college student might call her mother in the evening to discuss her day's studies. She is unlikely to call her mother every day at 6:00 p.m.; she might call at 6:14 p.m. one day and 5:54 p.m. the next. Thus, rather than exact temporal information, a time segment or interval, such as between 5:50 PM and 6:15 PM, is very informative to capture her activity patterns. An optimal segmentation technique is needed to generate such time segments that capture similar behavioral characteristics [33]. As with the discretization of temporal information, a method to pre-process continuous context values to convert them to nominal values before applying the rule mining technique should be developed.

- *Rule Discovery and Model Building:* The majority of current context-aware frameworks are based on static, centralized client-server architectures [34]. Mobile platforms, on the other hand, demand that the context modeling process and inference engine be simple and lightweight. The discovery and analysis of contextually responsive features and their patterns are of great importance to make context-aware intelligent decisions in a ubiquitous computing environment [28]. Because of the vast amount of data processing, conventional computational techniques such as data mining and machine learning [1] may not be applicable to make real-time decisions for cell phone users, reducing the efficiency of mobile phones. The association rule mining technique [35], for example, can uncover a large number of redundant rules that are no longer useful, rendering the decision-making process complex and ineffective. To achieve large-scale context awareness and create smart context-aware models, a deeper understanding of the strengths and limitations of state-of-the-art big data processing and analytics systems is needed. Thus the challenge is to model the contextual data effectively for adequate decision-making. Mining in-depth behavioral patterns of mobile phone users based on related contextual information, utilizing mobile phone data collected from single or multiple sources, can be considered. Using various constraints and particular mobile phone application interestingness into models could be useful in the dynamic context-aware decision-making process. For instance, extracted phone call response behavioral rules according to an individual's preferences can be used to minimize phone call interruptions.

- *Knowledge-Based Interactive Postmining:* Knowledge-based postmining of discovered rules may be another research topic in terms of semantically generalizing rules to prevent categorical data sparsity and make the rules more useful and interesting in a specific domain [30]. Semantic generalization, which makes use of applicable domain information, broadens the scope of the rules by examining their semantic relationships. The semantic generalization of rules is crucial for enhancing context-based adaptation and ensuring that individuals in context-aware applications and services behave properly. For several applications, it is difficult for decision-makers to process, interpret, and use the generated data relevant rules in the decision-making process. Furthermore, it is crucial to choose the best rules based on the query context by using fewer but closely related rules. Thus, two issues must be considered for successful use of the created rules: discovering the less closely related rules, and managing categorical data sparseness when implementing these rules [5]. The concept of knowledge-based multi-level generalization of generated rules may help to solve the problems described above. Generalized association rules can help to reduce the search space and combine several low support rules into a less number of high support rules by taking into account the use of a concept hierarchy for a particular domain. An ontology-based approach [24, 36] could be a possible way to use such concepts for a particular domain while generalizing the rules.
- *Mobility and Adaptation:* Computing environments are highly diverse due to the mobility of computing devices, applications, and people. Pervasive computing applications are exposed to changes in available resources such as network access, input and output devices, unlike desktop applications, which rely on a carefully designed and relatively static collection of resources [28]. Furthermore, to complete tasks on behalf of customers, they are often expected to collaborate randomly and opportunistically with previously unknown software services. As a result, ubiquitous computer applications must be extremely adaptable and versatile. As an example, an application may need to modify its style of output following a transition from an office environment to a moving vehicle, to be less intrusive. Thus the challenge is to adapt the changing environment more

effectively in the system. As the recent advances in smart mobile phones can capture the current contextual information about user's mobility and dynamic environment, an adaptive context-aware application based on such contexts could be useful, which can adapt the contexts and behave accordingly to assist the individual users in their daily activities.

10.3 Concluding Remarks

In this book, we have done a comprehensive study on context-aware machine learning modeling utilizing users' mobile phone data, which can create a vital turn in the way of interaction among people and mobile devices in our real-world life. A comprehensive survey on this topic through a context-aware machine learning framework that explores multi-dimensional contexts in machine learning modeling, discretization analysis and time-series modeling, contextual rule discovery and predictive analytics, and recent-pattern based behavior modeling, has been conducted to provide intelligent services. Furthermore, we have also discussed how the extracted contextual rules can play a vital role to build a mobile expert system as well as the importance of deep neural network learning methods in the area. The analysis and mobile data mining techniques detailed in this book provides the basis for further research into machine learning rule-based modeling, and the potential to use such rules to build smart context-aware mobile applications for the end mobile phone users to intelligently assist them in their daily activities in a pervasive computing environment. The prominent application fields of context-aware machine learning modeling are many, but not limited to, personalized mobile applications, recommendation systems, IoT applications, smart city and systems, as well as smart cybersecurity services discussed throughout the book.

Finally, numerous ideas for future research are proposed to broaden this area to more applicable and pervasive scenarios. There were also some notes on potential solutions and suggestions for researchers, as well as a variety of future challenges. Overall, we do believe that our study opens a promising path for future research on context-aware rule-based modeling based on machine learning techniques, and to build personalized rule-based smart and intelligent systems for the end mobile phone users to intelligently assist themselves in their daily life. We conclude this book with the hope that it will be useful in the development of context-aware machine learning which will lead to a brighter future in a variety of applications within the scope of the Fourth Industrial Revolution (Industry .40), especially in the domain of data-driven smart computing and decision-making intelligence in our real-life services.

References

1. Sarker, I. H. (2021). Machine learning: Algorithms, real-world applications and research directions. *SN Computer Science, 2*(3), 1–21.
2. Grosan, C., & Abraham, A. (2011). Rule-based expert systems. In *Intelligent systems* (pp. 149–185). Springer.
3. Chang, Y. J., & Tang, J. C. (2015, August). Investigating mobile users' ringer mode usage and attentiveness and responsiveness to communication. In *Proceedings of the 17th International Conference on Human-Computer Interaction with Mobile Devices and Services* (pp. 6–15).
4. Pejovic, V., & Musolesi, M. (2014, September). InterruptMe: designing intelligent prompting mechanisms for pervasive applications. In *Proceedings of the 2014 ACM International Joint Conference on Pervasive and Ubiquitous Computing* (pp. 897–908).
5. Sarker, I. H. (2019). Context-aware rule learning from smartphone data: survey, challenges and future directions. *Journal of Big Data, 6*(1), 1–25.
6. Mehrotra, A., Hendley, R., & Musolesi, M. (2016, September). PrefMiner: mining user's preferences for intelligent mobile notification management. In *Proceedings of the 2016 ACM International Joint Conference on Pervasive and Ubiquitous Computing* (pp. 1223–1234).
7. Kanjo, E., Kuss, D. J., & Ang, C. S. (2017). NotiMind: utilizing responses to smart phone notifications as affective sensors. *IEEE Access, 5*, 22023–22035.
8. Turner, L. D., Allen, S. M., & Whitaker, R. M. (2015). Push or delay? Decomposing smartphone notification response behaviour. In *Human behavior understanding* (pp. 69–83). Springer.
9. Bobadilla, J., Ortega, F., Hernando, A., & Gutiérrez, A. (2013). Recommender systems survey. *Knowledge-Based Systems, 46*, 109–132.
10. Shin, D., Lee, J. W., Yeon, J., & Lee, S. G. (2009, July). Context-aware recommendation by aggregating user context. In *2009 IEEE Conference on Commerce and Enterprise Computing* (pp. 423–430). IEEE.
11. Park, M. H., Hong, J. H., & Cho, S. B. (2007, July). Location-based recommendation system using bayesian user's preference model in mobile devices. In *International Conference on Ubiquitous Intelligence and Computing* (pp. 1130–1139). Springer.
12. Zheng, Y., Xie, X., & Ma, W. Y. (2010). Geolife: A collaborative social networking service among user, location and trajectory. *IEEE Data Engineering Bulletin, 33*(2), 32–39.
13. Liu, B., Kong, D., Cen, L., Gong, N. Z., Jin, H., & Xiong, H. (2015, February). Personalized mobile app recommendation: Reconciling app functionality and user privacy preference. In *Proceedings of the Eighth ACM International Conference on Web Search and Data Mining* (pp. 315–324).
14. Zhu, H., Chen, E., Xiong, H., Yu, K., Cao, H., & Tian, J. (2014). Mining mobile user preferences for personalized context-aware recommendation. *ACM Transactions on Intelligent Systems and Technology (TIST), 5*(4), 1–27.
15. Stefanis, V., Plessas, A., Komninos, A., & Garofalakis, J. (2014). Frequency and recency context for the management and retrieval of personal information on mobile devices. *Pervasive and Mobile Computing, 15*, 100–112.
16. Phithakkitnukoon, S., Dantu, R., Claxton, R., & Eagle, N. (2011). Behavior-based adaptive call predictor. *ACM Transactions on Autonomous and Adaptive Systems (TAAS), 6*(3), 1–28.
17. Plessas, A., Stefanis, V., Komninos, A., & Garofalakis, J. (2017). Field evaluation of context aware adaptive interfaces for efficient mobile contact retrieval. *Pervasive and Mobile Computing, 35*, 51–64.
18. Phithakkitnukoon, S., & Dantu, R. (2011). Towards ubiquitous computing with call prediction. *ACM SIGMOBILE Mobile Computing and Communications Review, 15*(1), 52–64.
19. Baeza-Yates, R., Jiang, D., Silvestri, F., & Harrison, B. (2015, February). Predicting the next app that you are going to use. In *Proceedings of the Eighth ACM International Conference on Web Search and Data Mining* (pp. 285–294).

20. Kim, J., & Mielikäinen, T. (2014, September). Conditional log-linear models for mobile application usage prediction. In *Joint European Conference on Machine Learning and Knowledge Discovery in Databases* (pp. 672–687). Springer.
21. Zhu, H., Cao, H., Chen, E., Xiong, H., & Tian, J. (2012, October). Exploiting enriched contextual information for mobile app classification. In *Proceedings of the 21st ACM International Conference on Information and Knowledge Management* (pp. 1617–1621).
22. Zhu, H., Chen, E., Xiong, H., Cao, H., & Tian, J. (2013). Mobile app classification with enriched contextual information. *IEEE Transactions on Mobile Computing, 13*(7), 1550–1563.
23. Luberg, A., Tammet, T. & Järv, P. (2011). Smart city: a rule-based tourist recommendation system. In *Information and Communication Technologies in Tourism 2011* (pp. 51–62). Springer.
24. Sarker, I. H., Furhad, M. H., & Nowrozy, R. (2021). AI-driven cybersecurity: an overview, security intelligence modeling and research directions. *SN Computer Science, 2*(3), 1–18.
25. Bandinelli, M., Paganelli, F., Vannuccini, G., & Giuli, D. (2009, June). A context-aware security framework for next generation mobile networks. In *International Conference on Security and Privacy in Mobile Information and Communication Systems* (pp. 134–147). Springer.
26. International Telecommunication Union. (2003). ITU-T X.805 Security architecture for systems providing end-to-end communications SERIES X: Data Networks and Open System Communications – Security.
27. Sarker, I.H. (2021). Data Science and Analytics: An Overview from Data-Driven Smart Computing, Decision-Making and Applications Perspective. *SN Computer Science, 2*(377), 1–22.
28. Sarker, I. H., Hoque, M. M., Uddin, M. K., & Alsanoosy, T. (2021). Mobile data science and intelligent apps: concepts, ai-based modeling and research directions. *Mobile Networks and Applications, 26*(1), 285–303.
29. Frénay, B., & Verleysen, M. (2013). Classification in the presence of label noise: a survey. *IEEE Transactions on Neural Networks and Learning Systems, 25*(5), 845–869.
30. Sarker, I. H. (2019). Research issues in mining user behavioral rules for context-aware intelligent mobile applications. *Iran Journal of Computer Science, 2*(1), 41–51.
31. Sarker, I. H. (2019). A machine learning based robust prediction model for real-life mobile phone data. *Internet of Things, 5*, 180–193.
32. Halvey, M., Keane, M. T., & Smyth, B. (2005, September). Time-based segmentation of log data for user navigation prediction in personalization. In *The 2005 IEEE/WIC/ACM International Conference on Web Intelligence (WI'05)* (pp. 636–640). IEEE.
33. Sarker, I. H., Colman, A., Kabir, M. A., & Han, J. (2018). Individualized time-series segmentation for mining mobile phone user behavior. *The Computer Journal, 61*(3), 349–368.
34. Nalepa, G. J., & Bobek, S. (2014). Rule-based solution for context-aware reasoning on mobile devices. *Computer Science and Information Systems, 11*(1), 171–193.
35. Agrawal, R., & Srikant, R. (1994, September). Fast algorithms for mining association rules. In *Proc. 20th Int. Conf. Very Large Data Bases, VLDB* (Vol. 1215, pp. 487–499).
36. Maedche, A., & Staab, S. (2001). Ontology learning for the semantic web. *IEEE Intelligent Systems, 16*(2), 72–79.

Printed in the United States
by Baker & Taylor Publisher Services